AF370316

Bridging the Gap: Local Models and Efficient Braiding for Fractional Quantum Hall on Lattices

Chopra

Copyright © 2024 by Chopra

All rights reserved. No part of this book may be reproduced in any man-ner whatsoever without written permission except in the case of brief quotations embodied in critical articles and reviews.
First Printing, 2024

Table of contents

Chapter 1

Introduction

Low dimensional, interacting quantum many-body systems host a wealth of interesting phenomena ranging from quantum phase transitions at zero temperature to topological order. The physics of such systems are often described by their ground states that are special and occupy a rather tiny fraction of the Hilbert space. These small set of low energy states have little entanglement compared to the states that live in the middle of the spectrum. In this context, the most sought after are the topologically ordered systems which are characterized by ground states with specific degeneracy on a manifold, long range entanglement and excitations that obey fractional statistics known as anyons [1]. The classic example is the fractional quantum Hall (FQH) effect, which is a phase induced when a 2D gas of interacting electrons is subject to large magnetic fields [2, 3].

Interestingly, there have been multiple proposals to realize fractional quantum Hall physics on lattices which have several advantages over conventional solid state systems. One of the main goals of studying FQH on lattice is also to explore methods to realize anyons and to successfully braid them. In this direction, conformal field theory (CFT) has been a very useful tool to construct analytical states on lattice that describe FQH phases [4, 5] and even anyons [6] on quite large systems. Parent Hamiltonians for these analytical states have been constructed that are few body but non-local [5]. Although there have been proposals to construct local versions of these Hamiltonians in some cases by relying on symmetries and optimization [7], a general scheme to achieve local models is lacking. When it comes to realizing anyons and braiding them in experimentally relevant lattice models, there are potential challenges that arise due to the large size and shape of the anyons relative to the size of the lattice, the finite size issues and effects due to the presence of edges. Hence, braiding anyons efficiently on modest lattices with open boundaries is also lacking.

Partly, the purpose of this book is to address the above mentioned timely questions on FQH and anyons in a lattice setting. We discuss this in the following chapters.

In Chapter 2, we begin by introducing briefly to the fractional quantum Hall effect and anyons from a lattice perspective and end by discussing the Haldane-Shastry model whose ground state is a 1D discretization of the Laughlin state.

In Chapter 7, we propose a truncation scheme to tame the non-local FQH lattice Hamiltonians constructed from CFT. Specifically, we show that one can arrive at local Hamiltonians more generally by neither resorting to any sort of optimization, nor relying on symmetries in the models. The truncation method applies independently from the choice of lattice and boundary conditions. In some cases, we demonstrate numerically through exact diagonalization that, indeed, the ground states of these local Hamiltonians stabilize the desired FQH physics. Specifically, we confirm this through excellent wavefunction overlaps with the exact states, approximate but correct ground state degeneracies on the torus and topological entanglement entropies close to the expected values. We also discuss the construction of local Hamiltonians in 1D and demonstrate numerically that their ground states have the correct universal critical properties as the exact models.

In Chapter 8, we discuss how we efficiently braid Abelian anyons on a modest sized lattice with open boundaries. We demonstrate this in well studied FQH lattice models by optimizing the potentials on different lattice sites to achieve squeezed Abelian anyons with near zero overlap. Next, we discuss how we perform the braiding by crawling them like snakes ensuring that they do not overlap throughout the process and achieve near correct exchange statistics for the anyons. We show computations by assuming the strict adiabatic limit and later relax this to get an estimate of the time needed to perform the braiding operation.

Lack of local order parameters make it difficult to characterize a topologically ordered phase and one must rely on different set of indicators. Athough a number of tools have been developed to detect topological phase transitions, they apply under different conditions and do not cover the full range of many-body models. In Chapter 9, we demonstrate how anyons serve as a probe to capture topological quantum phase transition in quite different concrete examples. We show that, by exploiting a simple property such as the charge of the anyon, one can capture the topological phase. We show that one can modify the Hamiltonian locally to trap the quasiparticles at well-defined positions in the ground state. The excess charge around the defect is expected to be equal to the charge of the anyon in the topological phase. When the two phases fail to support the similar sets of quasiparticles, a change is seen in the charge at the phase transition point. We show that, in contrast to conventional probes such as topological entanglement entropy, many-body Chern number, spectral flow and entanglement spectrum, this approach is numerically cheap and does not require a particular choice of boundary conditions.

All of the above discussions seem to indicate that most of the interesting physics occurs only near the ground state of a quantum many-body system. This is because, the states high up in the spectrum of a generic system correspond to high temperatures and conform to the eigenstate thermalization hypothesis (ETH) [8, 9]. However, an exceptional case of strong violation of ETH in all the eigenstates became evident in many-body localization (MBL) [10], which is a phase induced by strong disorder in interacting quantum many-body systems in 1D. The excited states of a MBL system behave very much like gapped quantum ground states opening interesting possibilities like spontaneous symmetry breaking and topological ordering even in eigenstates at finite energy densities. Symmetries are also known to influence the eigenstates of disordered systems [11]. While MBL is possible in models that have discrete symmetries, systems with non-Abelian continuous symmetries fail to many-body localize even with large disorder. In the context of disordered systems, there are several questions that remain unanswered, like, the consequences of emergent symmetries in a MBL system, existence of MBL in long ranged models, non-ergodic phases beyond MBL, etc.

While MBL is an example of strong ergodicity breaking, there has been a sudden surge of interest in constructing models that violate ergodicity weakly through the mechanism of quantum many-body scars [12]. Scarring is basically due to a set of measure zero eigenstates that violate ETH which live in the midst of ETH obeying eigenstates of a generic system. There is now a growing list of scarred models with more than one mechanism responsible for their existence. The scar states are essentially similar to quantum ground states and hence also bear interesting features like quantum criticality and topological order. While there are, but, few models that describe scars with non chiral topological order, scar models with chiral topological order is lacking.

In part, this book also aims to discuss interesting non-ergodic phases that emerge in disordered models with non-local interactions and in the presence of different sorts of symmetries. Further, this book adds to the scarce list of scarred models with chiral and non-chiral topological order. We discuss this in the following chapters.

In Chapter 3, we first discuss thermalization in isolated quantum-many-body systems and mechanisms that violate thermalization like MBL and quantum many-body scars.

In Chapter 4, we demonstrate the emergence of a weakly broken MBL phase in a model Hamiltonian with non-local interactions that possesses a quantum critical ground state with an emergent SU(2) symmetry. We discuss how we introduce disorder in the model and show that the highly excited states display a transition from a thermal to a glassy MBL phase. However, interestingly, we show that the ground state is exceptional, which is not localized or glassy. We argue that, this is due to the emergent SU(2) symmetry in the ground state.

Further, we confirm through the analytical knowledge of the ground state, for large system sizes, that, the ground state stabilizes the Luttinger liquid phase even at significant disorder strengths. We also discuss a simple prescription to lift the special ground state to the middle of the spectrum of glassy MBL states. We call this as weakly broken MBL phase due to the presence of a non-MBL eigenstate midst the spectrum of MBL states. We address this special state to be an "inverted quantum scar", which, in some sense is opposite of quantum many-body scars that lead to the weak violation of ETH.

In Chapter 5, we study the effects of disorder in the well known Haldane-Shastry model which is SU(2) invariant. We demonstrate that at maximum disorder, the model has long ranged interactions that do not decay as a simple power law. Through computations of mid spectrum entanglement entropy and level spacing statistics, we show that the model fails to many-body localize even at large disorder due to the SU(2) symmetry. We also argue that this phase is non thermal for the system sizes investigated.

In Chapter 6, we discuss the construction of models with topological quantum scars with chiral and non chiral orders. As a first construction, we exploit the existence of more than one operators built through methods from CFT that annihilate the scar states of interest. In this case, the scar state is a lattice Laughlin state with chiral topological order in 2D and a quantum critical state with Luttinger liquid properties in 1D. We also introduce a free parameter in our models that pins the scar state at any desired position in the spectrum. We introduce small disorder in our models that ensures that the spectrum is generic and thermal. We confirm the chiral topological order in the scar state in 2D by demonstrating that the state supports anyons with the right charge and statistics. As a second construction, we demonstrate a general strategy to turn classes of frustration free lattice models into similar classes containing quantum many-body scars within the bulk of their spectrum while preserving much or all of the original symmetry. We apply this strategy to a well known class of quantum dimer models on the kagome lattice with a large parameter space. We discuss that the properties of the resulting scar states like the entanglement entropy are analytically accessible. We also demonstrate that the non-scar states conform to ETH through numerical exact diagonalization studies on lattices of up to 60 sites.

Chapter 2

Topological phases on lattice in lower dimensions

Quantumness of a phase of matter is relevant at zero temperature and hence governed by the ground states of the many-body Hamiltonian describing the system. It was believed that the remarkable Landau's principles of symmetry breaking could explain every possible quantum phase transition at zero temperature. After the discovery of the quantum Hall effect, it became increasingly clear that there exist phases of matter in 2D which simply do not fit under the Landau paradigm. These are the topologically ordered phases [13] whose complete classification has remained largely unsettled. Topological order can be classified broadly into symmetry protected topological phases (SPT) and those with intrinsic topological order. SPT phases are states of matter that have a finite energy gap and a particular symmetry protecting the phase. These are trivial topological phases characterised by entanglement which is short ranged. Examples include integer quantum Hall systems, topological insulators, etc. Intrinsic topologically ordered phases are characterised by long range entanglement. The most important features of intrinsic topological order are ground state degeneracy on a specific manifold and fractionalized excitations which are both absent in SPT phases.

The paradigmatic example for intrinsic topological order is the Fractional Quantum Hall Effect (FQHE) where electrons form an incompressible fluid in two dimensions when subject to a large magnetic field. In 1982, it occurred as a great surprise when Tsui, Stormer and Gossard found plateaux in the Hall resistivity at non-integer filling fractions [3]. Venturing the problem theoretically is extremely hard since it is unclear how electron interactions would lift the macroscopic degeneracy in a partially filled Landau level and invoking the standard degenerate perturbation theory would be futile. Then came the incredible proposal by Laughlin for the ground state wavefunction at filling fraction $1/m$ given by [2],

$$\Psi_m(z) = \prod_{i<j} (z_i - z_j)^m e^{-\sum_{i=1}^{N} |z_i|^2 / 4l_B^2} \tag{2.1}$$

. In the above expression, z_i are the positions of the electrons and l_B is the magnetic length. The wavefunction takes into account the antisymmetrization for electrons when m is odd and vanishes with a mth order zero when electrons approach each other, while the exponential factor decreases soon when the electrons move away from the center. This creates a perfect balance which qualifies the state to describe an incompressible fluid. Several numerical studies have indeed confirmed that Laughlin's wavefunction has an excellent overlap ($> 99\%$) with the ground state of electrons interacting via realistic Coulomb interactions.

An exact parent Hamiltonian whose zero energy ground state is the Laughlin state described in Eq. 2.1, can be written in terms of the Haldane pseudopotentials as below [14],

$$H = \sum_{m'=0}^{\infty} \sum_{i<j} v_{m'} P_{m'}(ij), \tag{2.2}$$

where, $P_m(ij)$ is the projection operator which chooses out states where particles i and j have a relative angular momentum m. If one considers a hardcore potential such as $v'_m = 0$ for $m' \geq m$, then we have $P_{m'}(ij)\psi_m(z) = 0$ for any $m' < m$. Then it immediately follows that, $H\psi_m(z) = 0$.

Laughlin's phenomenological guess works well only for filling fractions $v = 1/m$ where m is odd and there are now mechanisms proposed to capture other filling fractions in a single frame work [15].

The prime purpose of this chapter is to discuss the aspects of fractional quantum Hall physics on a lattice to set the stage for rest of the discussions in this book. In the following sections, we mainly discuss different routes to engineer FQH physics on a lattice, introduce briefly to anyons and end with a discussion on the famous Haldane-Shastry model that we repeatedly encounter in this book.

2.1 Lattice fractional quantum Hall effect

Realizing fractional quantum Hall (FQH) physics on lattice could be interesting in its own right due to the following reasons. Firstly, a natural question arises if the physics of FQH that may arise say through an interacting Bosonic system on a lattice is different than the continuum due to an underlying lattice. It has been shown that one can ignore lattice effects in the limit of small number of flux quanta per lattice plaquette [16, 17] and when this is not true, the presence of lattice can lead to new correlated states of matter that are absent in the continuum [18–20]. Secondly, lattice has several advantages experimentally and a number of proposals suggest that the FQH regime can be realized using optical lattices [21, 22, 17, 23, 24]. The prime advantage in a lattice is the possibility of achieving a stable and robust ground state separated by a larger gap to the excitations. Also, the need for large magnetic fields is not necessary on a lattice as compared to the continuum. Thirdly, there is much interest in creating and manipulating anyons from the view point of topological quantum computation [25] and lattice FQH systems show a wide range of possibilities in this direction as we will discuss further in this book.

We will discuss below the different routes through which a FQH phase on a lattice can be engineered.

2.1.1 Fractional Chern insulators

In continuum FQH, all Landau levels have the same Chern number and are dispersionless (flat) in the absence of disorder. Since the Landau levels are essentially flat, they can accommodate an exponentially large number of Fermion slater determinants when they are partially filled. It is from these macroscopically degenerate states, incompressible liquids are selected by interactions at special values of the filling fractions. Hence, to realize FQH physics in lattice, one must engineer flat Chern bands to mimic Landau levels and then add interactions [26–29]. These are dubbed as "fractional Chern insulators". It was demonstrated for instance in Ref. [29], where they start out with a local Bloch Hamiltonian H_k that has dispersing Chern bands and then flatten the bands by tuning the hopping parameters and thereby adding interactions. They demonstrate numerically through exact diagonalization studies that the ground state is gapped and stabilizes FQH order.

2.1.2 FQH on lattice through magnetic fields

Optical lattice setups are promising where one can engineer artificial guage fields to induce FQH physics and also allowing for detailed investigations of the effect, even at the level of single particles. The simplest such model is the Bose-Hubbard model in a magnetic field on a square lattice, where a bosonic $v = 1/2$ FQH phase was theoretically studied [16, 17]. The Hamiltonian of this model reads as,

$$H_{\text{BH}} = -t \sum_{xy} (\hat{c}^{\dagger}_{x+1,y} \hat{c}_{x,y} e^{-i\pi\alpha y} + \hat{c}^{\dagger}_{x,y+1} \hat{c}_{x,y} e^{i\pi\alpha x} + \text{h.c}) + U \sum_{x} \hat{n}_{x,y}(\hat{n}_{x,y} - 1), \qquad (2.3)$$

where $\hat{c}^{\dagger}_{x,y}(\hat{c}_{x,y})$ is the bosonic creation (annihilation) operator at position (x,y), t is the hopping amplitude, $U > 0$ is the interaction strength and $2\pi\alpha$ is the flux through the plaquette. When α is small and the number of particles per lattice site is small, this is essentially the continuum limit and the ground state of the model is the Laughlin's wavefunction described in Eq.(2.1). Numerical computations of the overlap between the Laughlin states on the torus and the lattice many-body ground states were close to unity when α was much smaller than the flux quantum. This overlap however decreased when α becomes of the order of one quarter of the flux quantum [17]. Further, it was shown through the computations of many-body Chern number that indeed there is a phase transition from a topological phase to a trivial phase in the above model when one changes the parameter α [16].

In Chapter 9, we study this phase transition through the insertion of defects that turn out to be anyons of the correct charge only in the topological phase. We also show to what extent the model supports topological order with disorder through the eye of anyons.

2.1.3   Conformal field theory construction

The idea of Moore and Read [30] that the FQH states can be expressed as conformal field theory (CFT) correlators can also be applied to construct lattice versions of fractional quantum Hall states. In general, the wavefunction on a lattice of N sites with a fixed particle number $\sum_i n_i = M$, where $n_j \in \{0,1\}$ is the number of particles on the jth site, has the form,

$$|\Psi\rangle = \frac{1}{C} \sum_{n_1,\dots,n_N} \psi(n_1,\dots,n_N)|n_1,\dots,n_N\rangle \ \text{ with } C^2 = \sum_{n_1,\dots,n_N} |\psi(n_1,\dots,n_N)|^2. \qquad (2.4)$$

The Laughlin wavefunction at the Landau level filling fraction $\nu_{LL} = 1/q$ on the lattice can be obtained through the correlator of operators from CFT as [5],

$$\psi(n_1,\dots,n_N) = \langle V_{n_1}(z_1)...V_{n_N}(z_N)\rangle. \qquad (2.5)$$

In the above expression, $V_{n_i}(z_i)$ is the vertex operator defined at position z_i given by,

$$V_{n_i}(z_i) =: e^{i(qn_i-\eta)\phi(z_i)/\sqrt{q}} : \qquad (2.6)$$

where, $\phi(z)$ is a chiral bosonic field of the $c = 1$ free bosonic CFT and $: \cdots :$ denotes normal ordering. By the suitable choice of the vertex operators, the charge neutrality condition is addressed. The correlator in (2.5) leads to the following wavefunction,

$$\psi_q^\eta(n_1,...,n_N) = \delta_n \prod_{i<k}(z_i - z_k)^{q n_i n_k} \prod_{j\neq l}(z_l - z_j)^{-\eta n_l}, \tag{2.7}$$

where $\delta_n = 1$ if the number of particles is $M = \frac{\eta N}{q}$ and $\delta_n = 0$ otherwise. Hence now, one can also define a lattice filling fraction $\nu_{\mathrm{LA}} = \frac{M}{N} = \frac{\eta}{q}$. When $\eta = 1$, we are in the strict lattice limit and we reach the continuum limit through $\eta \to 0^+$, $N \to \infty$ and by fixing ηN. The wavefunction in Eq. (2.7) describes fermions and bosons for q being odd and even respectively with the first product term interpreted as the attachment of q fluxes to each particle and the second product term corresponds to the background charge in the lattice.

One can also construct parent Hamiltonians for these states by exploiting the existence of null fields in CFT [5]. The approach to derive exact lattice Hamiltonians begins by finding a set of operators Λ_i which annihilate the lattice FQH state $|\Psi_{\mathrm{FQH}}\rangle$, i.e.,

$$\Lambda_i|\Psi_{\mathrm{FQH}}\rangle = 0 \tag{2.8}$$

The Hamiltonian, whose ground state is $|\Psi_{\mathrm{FQH}}\rangle$, is given by the following semidefinite operator,

$$H = \sum_i \Lambda_i^\dagger \Lambda_i \tag{2.9}$$

These parent Hamiltonians are few-body with nonlocal terms and in some cases, truncation and optimizing the coefficients have lead to local Hamiltonians with similar topological properties [31, 32]. In Chapter 7, we discuss a general truncation scheme to arrive at local FQH Hamiltonians without resorting to any sort of fine tuning. We show that the local Hamiltonians indeed stabilize the fractional quantum Hall order in their ground states. We also discuss how we obtain local Hamiltonians in 1D whose ground states retain the correct universal critical properties.

2.2 Anyons

Quantum statistics is a crucial ingredient for quantum mechanical structure of the microscopic world. It is the defining property which states that the wavefunction describing a system of identical particles should satisfy the proper symmetry under the exchange of any two particles. For instance, fermions and bosons are distinguished from the requirement that under exchange, the wavefunction of bosons is symmetric while that of fermions is antisymmetric. Furthermore, in three spatial dimensions, moving one particle all the way around an other is topologically equivalent to a state where the particles have not moved at all. Hence the only two possibilities are the wavefunctions that can utmost change by a $\pm$ sign under exchange leading to the possibility of existence of just bosons and fermions in three dimensions. However, this is not the case in two spatial dimensions where the particle loop that encircles the other particle cannot be deformed to a point continuously. Hence exchanging a particle with the other in two spatial dimensions may lead to an arbitrary phase,

$$\psi(\mathbf{r}_1,\mathbf{r}_2) \to e^{i\theta}\psi(\mathbf{r}_1,\mathbf{r}_2) \tag{2.10}$$

Here the phase need not just be a $\pm$ sign and can be arbitrary since exchanging the particle again in the same direction can lead to a non trivial phase :

$$\psi(\mathbf{r}_1,\mathbf{r}_2) \to e^{2i\theta}\psi(\mathbf{r}_1,\mathbf{r}_2) \tag{2.11}$$

Hence bosons and fermions are just special cases when $\theta = 0, \pi$. Anyons are particles with any statistical phase θ [33]. Anyons described above are of the Abelian kind where different exchanges lead to phases that commute. However, there can be an exotic possibility when the anyons are non Abelian, where degenerate ground states mix among themselves under an exchange operation. Realizing non-Abelian anyons experimentally and braiding them could be a crucial first step towards constructing a fault tolerant topological quantum computer [25].

The best example where anyons show up is again the fractional quantum Hall effect where excitations are quasi-holes and quasi-electrons. It is well-known how one can modify the Laughlin state [2] with quasiholes defined on a disk-shaped region in the two-dimensional plane into a Laughlin state with quasiholes defined on a square lattice with open boundary conditions [6]. This is done by restricting the allowed particle positions to the lattice sites and by also restricting the magnetic field to only go through the lattice sites. Here, we consider a system with one magnetic flux unit through each lattice site, and take the charge of a particle to be -1. The resulting wavefunction with S quasiholes at the positions w_i, with $i = 1, 2, \ldots, S$, can again be constructed through a CFT correlator [6]. It is given by,

$$|\Psi_{q,S}\rangle = \sum_{n_1,n_2,\ldots,n_N} \Psi_{q,S}(n_1,n_2,\ldots,n_N)|n_1,n_2,\ldots,n_N\rangle, \tag{2.12}$$

where

$$\Psi_{q,S}(n_1,n_2,\ldots,n_N) = C^{-1}\delta_n \prod_{j,k}(w_j - z_k)^{p_j n_k} \times \prod_{i<j}(z_i - z_j)^{q n_i n_j} \prod_{i \neq j}(z_i - z_j)^{-n_i}. \tag{2.13}$$

In this expression, the z_i, with $i = 1, 2, \ldots, N$, are the coordinates of the N lattice sites written as complex numbers, $n_j \in \{0, 1\}$ is the number of particles on the jth site, q and p_j are not too large integers, and q must be at least 2. C is a real normalization constant that depends on all the w_i, and

$$\delta_n = \begin{cases} 1 & \text{for } \sum_{j=1}^{N} n_j = \left(N - \sum_{j=1}^{S} p_j\right)/q \\ 0 & \text{otherwise} \end{cases} \tag{2.14}$$

fixes the number of particles $\sum_j n_j \gg 1$ in the system in such a way that there are q flux units per particle and p_k flux units per quasihole. The particles are fermions for q odd and hardcore bosons for q even. We restrict q and p_j to small integers and require the number of particles to be large compared to one, since this is the regime for which it has already been confirmed numerically [6, 34] that the lattice Laughlin state has the desired topological properties. When the state is topological, one observes the following results numerically [6, 34]. The state without anyons has a uniform density of $1/q$ particles per site in the bulk of the system, which means that $\langle \psi_{q,0} | n_j | \psi_{q,0} \rangle$ is constant and equal to $1/q$ in the bulk. The kth quasihole creates a local region around w_k with a lower particle density. This can be quantified by considering the density difference

$$\rho(z_j) = \langle \psi_{q,s} | n_j | \psi_{q,s} \rangle - \langle \psi_{q,0} | n_j | \psi_{q,0} \rangle, \tag{2.15}$$

which is defined as the expectation value of n_j in the state with anyons minus the expectation value of n_j in the state without anyons. When all other anyons are far away, the total number of particles missing in the local region is p_k/q. This must be so when the quasihole is entirely inside the local region, since it follows from Eq. (2.14) that the presence of the kth quasihole reduces the number of particles in the system by p_k/q. Since the particles have charge -1, the quantity $-\rho(z_j)$ is the charge distribution of the anyons, and summing $-\rho(z_j)$ over the local region at w_k gives the charge p_k/q of the kth anyon.

Creation of localized anyons through pinning potentials added to the FQH Hamiltonians have also been studied [35, 36]. Although there is yet no direct observation of fractional statistics, some signatures have been observed in electron interferometer experiments [37, 38]. As discussed earlier, in a lattice, there is now much interest to engineer FQH physics and hence naturally also to realize anyons. Experimentally, optical and atomic systems have an advantage over conventional solid-state systems since there is a possibility to create and control quasi particles more naturally and with precision [39]. Similar to the continuum, fractional excitations can be invoked through trapping potentials added to the lattice FQH Hamiltonians [40–44]. Braiding experiments have been performed numerically by adiabatically moving the trapping potentials to move the anyons [40]. However, on a finite sized system with open boundaries and with few lattice sites, the anyons typically have a spread that lead to a significant overlap and hence braiding results are not very satisfactory. In Chapter 8, we address this issue by optimizing the trapping potentials so as to squeeze the anyons on few sites on a modest lattice with open boundaries. Interestingly, we show that, in this way, anyons can be crawled like a snake with minimum overlap and numerically achieve far better braiding results.

2.3　The Haldane-Shastry model

Interestingly, if one restricts attention to one dimension there exists a model spin Hamiltonian whose ground state resembles a 1D discretization of the Laughlin state. This is the famous " Haldane-Shastry model (HS model)" [45, 46]. The model is a Heisenberg antiferromagnetic spin chain with couplings that decay as the square of the inverse distance between the spins located on the unit circle (at positions $z_j = \exp{(i\frac{2\pi j}{N})}$) given by,

$$H_{\text{HS}} - \left(\frac{2\pi}{N}\right)^2 \sum_{i<j} \frac{\mathbf{S_i}.\mathbf{S_j}}{|z_i - z_j|^2} \tag{2.16}$$

where $|z_i - z_j|$ is the chord distance between the sites i and j, as depicted in the Fig. 2.1. The Hamiltonian is SU(2) invariant ($[H^{\text{HS}}, \mathbf{S}_{\text{tot}}] = 0$) and has an additional symmetry generated by the rapidity operator given by,

$$\Lambda = \frac{i}{2} \sum_{i,j=1 i\neq j}^{N} \frac{z_i + z_j}{z_i - z_j} \mathbf{S_i} \times \mathbf{S_j}, \qquad [H_{\text{HS}}, \Lambda] = 0. \tag{2.17}$$

Although both $\mathbf{S}_{\text{tot}}$ and Λ commute with the Hamiltonian, they do not commute with each other and this generates an infinite dimensional Yangian algebra making the model completely solvable [47–50]. The ground state of this model is known exactly and given by,

$$|\psi_{\text{HS}}\rangle = \sum_{z_1,...,z_M} \psi_{\text{HS}}(z_1,...,z_M) \, S_{z_1}^+...S_{z_M}^+ |\downarrow\downarrow .....\downarrow\rangle. \tag{2.18}$$

In the above expression, the sum runs over all possible ways to distribute the coordinates z_i of $M = \frac{N}{2} \uparrow$ spin on the unit circle. The amplitude is given by,

$$\psi_{\text{HS}}(z_1,...,z_M) = \prod_{i<j}^{M} (z_i - z_j)^2 \prod_{i=1}^{M} z_i. \tag{2.19}$$

As emphasized in the beginning, the above wavefunction is remarkably similar to the Bosonic Laughlin's wavefunction at filling fraction $1/2$ on a ring when one transforms the spin

Fig. 2.1 System of N spin $1/2$ particles on the unit circle at positions $z_j = \exp\left(i\frac{2\pi j}{N}\right)$. The chord distance $|z_i - z_j|$ is shown with the straight line connecting the spins at positions i and j.

variables to hardcore bosons. In fact, the wavefunction expressed in Eq. (2.7) corresponds to the lattice HS wavefunction for $q = 2$ and $\eta = 1$ in the language of hardcore bosons. The excitations of the model are spinons and can be constructed in a way similar to the construction of quasiholes starting from the Laughlin's wavefunction [51–53]. In addition, this model is known to be the conformal gapless fixed point of the spin models within the $SU(2)_1$ WZW class, a Luttinger liquid [45, 54, 55].

The HS model requires the constraint that the spins be positioned uniformly on the unit circle. It would be interesting to ask how the physics of the model and its ground state get affected by relaxing such a constraint. In Chapter 4 we will address this by studying the effects of disorder in a non $SU(2)$ invariant parent Hamiltonian for an inhomogeneous version of the HS wavefunction. Interestingly, we find that the state stabilizes Luttinger liquid phase even with a fair amount of disorder. In Chapter 5, we will discuss the effects of disorder in an inhomogeneous $SU(2)$ version of the HS model and show numerically an emergence of a non-ergodic but also a non-MBL phase at maximum disorder.

Chapter 3

Thermalization and its violation in quantum many-body systems

The concept of thermalization has successfully captured the emergent macroscopic behaviour of physical systems. However, a description at the microscopic level has remained elusive and still being debated. In classical mechanics, non-linear equations governing the particle trajectories drive them to explore the constant-energy space uniformly according to the microcanonical measure. However, for an isolated quantum many-body system, dynamics is inherently unitary and hence needs a different treatment to understand the process of thermalization. On the other hand, a natural question arises if every generic quantum many-body system void of special conservation laws always thermalizes or do routes exist to escape thermalization.

In this chapter, we will introduce briefly to the concept of thermalization in isolated quantum systems. Next, we describe how an interplay between disorder and interactions can lead to a strong violation of thermalization in the many-body localized systems. We will also briefly discuss the consequence of long ranged terms and symmetries on the physics of many-body localization. We will end by discussing the phenomenon of quantum many-body scars which is yet an other mechanism through which thermalization is avoided in the weaker sense.

3.1 Thermalization in isolated quantum many-body systems

The breakthrough in understanding thermalization in quantum systems came with the formulation of "eigenstate thermalization", where a single eigenstate can be thought of being in thermal equilibrium. An ansatz was put forth through the name of "Eigenstate Thermalization Hypothesis (ETH)" which gave a solid foundation to the mechanism of eigenstate thermalization [8, 9]. The hypothesis comes with the description for matrix elements of a local observable $\hat{O}$ computed within the eigenstates $|\alpha\rangle$ and $|\beta\rangle$ with the corresponding energies E_α and E_β of a generic, non-integrable quantum many-body system given by,

$$\langle\alpha|\hat{O}|\beta\rangle = O_{\mathrm{mc}}(\overline{E})\delta_{\alpha\beta} + \exp\left(-S_{\overline{E}}/2\right)R_{\alpha\beta}f(\omega,\overline{E}) \tag{3.1}$$

where, $\overline{E} = (E_\alpha + E_\beta)/2$, $\omega = E_\alpha - E_\beta$, $S_{\overline{E}}$ is the thermodynamic entropy which scales as the volume of the system ($S \propto V$), O_{mc} is the microcanonical average and $R_{\alpha\beta}$ are independent normally distributed random variables. In addition, $O_{\mathrm{mc}}(\overline{E})$ and $f(\omega,\overline{E})$ are required to be smooth functions of their arguments.

Now, if one considers a non-equilibrium initial state $|\psi(0)\rangle$ and computes the average of any local observable $\hat{O}$ in the time evolved state after sufficiently long time, one would get an equilibrium expectation value as shown below.

$$|\psi(t)\rangle = \sum_n C_n e^{-iE_n t}|E_n\rangle \tag{3.2}$$

where, $|E_n\rangle$ are eigenstates of a generic quantum many-body Hamiltonian H and $C_n = \langle E_n|\psi(0)\rangle$. The expectation value of the local operator $\hat{O}$ in the time evolved state $|\psi(t)\rangle$ is given by,

$$\langle\hat{O}(t)\rangle = \langle\psi(t)|\hat{O}|\psi(t)\rangle = \sum_n |C_n|^2 + \sum_{n\neq m} C_m^* C_n e^{-i(E_n-E_m)t} O_{mn} \tag{3.3}$$

where, $O_{mn} = \langle E_m|\hat{O}|E_n\rangle$. Long time average of the observable in the absence of symmetries and degeneracies is given by,

$$\lim_{T'\to\infty} \frac{1}{T'} \int_0^{T'} \langle\hat{O}(t)\rangle = \sum_n |C_n|^2 O_{nn} \tag{3.4}$$

Now, if the system thermalizes, one can invoke ETH to get,

$$\sum_n |C_n|^2 O_{nn} \approx O_{\mathrm{mc}}. \tag{3.5}$$

3.1.1 Numerical evidences for thermalization

There have been several numerical checks for eigenstate thermalization hypothesis. As an example, we report the studies in the well known interacting model of hardcore bosons from the Ref [56] given by

$$H = \sum_{i=1}^{N}[-J(c_i^\dagger c_{i+1}+\text{H.C})+V(\hat{n}_i-\tfrac{1}{2})(\hat{n}_{i+1}-\tfrac{1}{2})-J'(c_i^\dagger c_{i+2}+\text{H.C})+V'(\hat{n}_i-\tfrac{1}{2})(\hat{n}_{i+2}-\tfrac{1}{2})]$$

$$(3.6)$$

In the above Hamiltonian, $c_i^\dagger$ (c_i) is the creation (annihilation) operator for the hardcore boson at the lattice site i and $\hat{n}_i = c_i^\dagger c_i$ is the number operator at the lattice site i. When $J' = V' = 0$, the model maps on to the well known spin-1/2 XXZ spin chain which is integrable. The model becomes generic and quantum chaotic when it is tuned away from integrability and the numerical evidence is discussed below.

3.1.1.1 Level spacing statistics

Indications of thermalization can be inferred by studying the statistics of the energy level spacing of a many-body system. The essential idea is that the Hamiltonian of a generic system is in fact a random matrix and hence there is a tendency for energy level repulsion. Therefore, the distribution of spacing s between consecutive energy levels would be that of Wigner-Dyson. In particular, if one considers real symmetric Hamiltonians, one would get a distribution that belongs to the Gaussian Orthogonal Ensemble (GOE) given by $P_{\text{GOE}}(s) = \frac{\pi}{2}se^{-\frac{\pi}{4}s^2}$ and for complex hermitian Hamiltonians, one would get Gaussian Unitary Ensemble (GUE) distribution given by $P_{\text{GUE}}(s) = \frac{32}{\pi^2}s^2e^{-\frac{4}{\pi}s^2}$. On the other hand, for systems that are integrable, there is an absence of level repulsion and one would get Poissonian distribution for the level spacing given by $P(s) = e^{-s}$.

It is important to carry out the analysis of level spacing distribution in the sectors of Hamiltonians free from any symmetries and failing to do so can cause the quantum chaotic system to appear integrable. This is because of the absence of level repulsion between levels in different symmetry sectors. The model described in Eq. 3.1.1 is translationally invariant and different quasi-momentum sectors labelled by k are decoupled. The distribution of the level spacing ω, $P(\omega)$, was numerically computed by averaging over all the k-sectors after fixing the number of particles. At the integrable point ($J' = V' = 0$), $P(\omega)$ was shown to be close to the Poisson distribution and for greater values of the integrability breaking strengths, $P(\omega)$ became close to GOE distribution [56].

3.1.1.2 Entanglement entropy

The entanglement structure of the eigenstates have direct implications for an ergodic system. For an eigenstate $|r\rangle$ of an ergodic Hamiltonian that obeys ETH, we saw that all local observables in a subsystem will have thermal expectation values. This means that the reduced density matrix of a subsystem A after tracing the rest of the system B, $\rho_A = \text{Tr}_B|r\rangle\langle r|$ is thermal. Therefore, the von Neumann entropy of ρ_A which measures the amount of entanglement in the state $|r\rangle$ should be equal to the thermodynamic entropy. That is,

$$S_{\text{ent}}(A) = -\text{Tr}\,\rho_A \ln(\rho_A) = S_{\text{th}}(A). \tag{3.7}$$

The thermodynamic entropy is an extensive quantity and hence one would expect that entanglement entropy of the states in the middle of the spectrum obeys "volume-law" scaling proportional to the volume of the subsystem, $S_{\text{ent}}(A) \propto \text{vol}(A)$. Random matrix theory predicts that the entropy of a high energy eigenstate of an ergodic system should be equal to the "Page value" given by, $S_{\text{Page}} = \ln(0.48D) + O(1/D)$, where D is the dimension of the many-body Hilbert space [57]. Numerical findings indicated that the entropy S_m for eigenstates of the Hamiltonian in Eq. 3.1.1 are only bounded by the Page value [56].

3.2 Routes to break ergodicity in quantum many-body systems

As we saw, a generic system which does not have special conservation laws thermalizes. The question naturally arises if there are mechanisms through which a system avoids to reach thermal equilibrium even after long enough times. Integrable systems are known to possess an extensive number of integrals of motion and hence do not thermalize in the usual sense. However, integrability is a fine tuned phase and any perturbation will usually make the system generic. Two well known examples of ergodicity breaking in generic systems are many-body localization and quantum many-body scars as we discuss them below.

3.2.1 Many-body localization

The physics of a single particle on a lattice is described by a propagating Bloch state similar to plane waves. Naively one would think that in the presence of some disorder, quantum mechanical tunnelling process would still permit for the diffusion process to occur. However, Anderson in his seminal paper showed that the wavefunction of the single particle in a disordered environment becomes localized in a certain region of space and decays exponentially away from that region [58]. The seed of localization lies in the fact that at extreme disorder, the variance of the disorder is much large compared to the tunnelling amplitude between neighbouring lattice sites which makes the resonant transitions impossible. In dimensions three or greater, for weak disorder, the eigenstates of the Hamiltonian are extended and the dynamics of the particle is diffusive. However, in dimensions one and two, even for arbitrarily weak disorder, each eigenstate labelled by α is exponentially localized near position R_α with wavefunctions that have the long distance asymptotic form given by $\psi_\alpha(\vec{r}) \sim \exp(\frac{-|\vec{r}-\vec{R}_\alpha|}{\xi})$. The localization length ξ typically depends on the disorder strength and the energy density. Thus the eigenstates clearly do not obey ETH and if one were to intialize the system with a non-uniform density pattern which corresponds to high energy, the system would retain the memory of the pattern even after long times.

The question thus arises if localization persists when interaction between particles is included. In a significant work by Basko, Aleiner and Altshuler (BAA) [59], it was shown that indeed Anderson localization is perturbatively stable to interactions. This phase was dubbed the many-body localized phase (MBL). The standard model that exhibits an MBL phase is the Heisenberg spin chain in the presence of random onsite magnetic field (h_i) [60] whose Hamiltonian reads as,

$$H = J \sum_{<i,j>} \vec{\sigma}_i.\vec{\sigma}_j + \sum_i h_i \sigma_i^z. \tag{3.8}$$

In the above expression, h_i is drawn from a uniform distribution of width $W > 0$. When $J = 0$, the system is trivially localized where all the eigenstates are product states which is of the form $|\sigma_1^z\rangle \otimes |\sigma_2^z\rangle .. \otimes |\sigma_N^z\rangle$. For $J \ll W$, the many-body eigenstates can be perturbatively constructed starting from the product state. In this limit, the energy level difference between different sites is larger than the interaction strength J and thus the states on different lattice sites weakly hybridize. Thus, thermalization does not occur and there is absence of transport.

There is a phenomenological treatment to MBL which is described as follows [61]. For simplicity, if we consider a system of N spin 1/2 degrees of freedom governed by a Hamiltonian such as the one described in (3.8), at large disorder, all of the eigenstates can be localized. In this limit, one can construct new localized two level degrees of freedom called the '1-bits' denoted by $\vec{\tau}_i$. These 1-bits are constructed through quasi-local unitary transformations ($\vec{\tau} = U^\dagger \vec{\sigma} U$) and hence the 1-bits are close to the original spin operators at least at very large disorder. At large disorder, the Hamiltonian can now be rewritten in terms of the $\vec{\tau}$ operators as,

$$H = \sum_i \tau_i^z + \sum_{i>j} J_{ij} \tau_i^z \tau_j^z + \sum_{i>j>k} J_{ijk} \tau_i^z \tau_j^z \tau_k^z + ... \tag{3.9}$$

Since the above Hamiltonian was derived through quasi-local unitary transformations, the couplings decay exponentially with separation between the τ^z operators. Also since $[H, \tau_i^z] = 0$, the eigenstates of H can be fully specified by labelling them with the eigenvalues of all τ_i^z. Hence, one can view each τ_i^z operator as an emergent conserved quantity which cannot decay during the quantum time evolution. Thus there is a sense of emergent integrability in MBL systems due to the existence of the complete set of local conserved quantities. This is the reason why the energy level spacings obey the Poissonian statistics in a MBL system. However, in contrast to the fine tuned integrable systems, integrability in MBL systems is robust to perturbations which qualifies it as a generic example of ETH violation.

An other important feature of MBL systems is the low entanglement in the eigenstates, typically area law entanglement [10]. The entanglement entropy of a subsystem A computed for an MBL eigenstate scales as the volume of the boundary ∂A of A given by $S \propto \text{vol}(\partial A)$, when both the size of the system and subsystem are taken to infinity. For a one dimensional system, ∂A is just two end points of the subsystem and hence area law implies that entanglement entropy does not scale with the size of subsystem, $S = \text{constant}$. An intuitive way to understand this is, suppose we consider two disconnected pieces A and B of a system with corresponding Hamiltonians H_A and H_B, then the eigenstates of the net system ($H_A + H_B$) is a simple tensor product $|AB\rangle = |A\rangle \otimes |B\rangle$ where $|A\rangle$ and $|B\rangle$ are eigenstates of H_A and H_B respectively. Now, if we turn on a local coupling near the boundary, in a MBL system, introducing a local perturbation will only affect degrees of freedom situated within the localization length ξ from the boundary. Thus the new eigenstates can be obtained from $|AB\rangle$ by entangling degrees of freedom in A and B within the distance ξ away from the boundary leading to an area law scaling of entanglement.

3.2.1.1   MBL in the presence of long ranged interactions and hopping

Intuitively, long ranged interactions and hopping processes in even largely disordered systems support the formation of an effective bath that can lead to thermalization and hence not favour MBL. Perturbative arguments suggest that, with the presence of power-law interactions and hopping terms $v, t \sim 1/r^\alpha$, MBL cannot survive when $\alpha < 2d$ where d is the dimensionality of the space [62–64]. The potential cause for a possible breakdown of MBL phase are resonance processes that lead to an eventual delocalization. In fact, numerical studies have shown that long ranged hopping can be more harmful for an MBL phase compared to long ranged interactions due to the divergence in the number of resonances in the case of long ranged hopping. Also, an intriguing possibility of MBL with long ranged interactions has been proposed in one dimensional systems where charge confinement is favoured due to interactions [65].

In Chapter 4, we show the existence of MBL phase in a model with non-local interactions and hopping terms. Although the hopping terms decay quickly with distance, the interaction terms do not have a simple power law decay. In fact, the particles that are farther apart can interact more strongly than the ones that are close by. This hence serves as a counter intuitive situation where MBL phase is stable although the model is sufficiently non-local.

3.2.1.2 Role of symmetries in MBL physics

The eigenstates of a many-body localized system have very low entanglement similar to the ground states of gapped system. It is well known that spontaneous symmetry breaking at finite temperature in one-dimensional systems with short range interactions is absent [66]. However, in an MBL system, the eigenstates violate ETH and hence can exhibit ordered phases in contrast to thermal ensembles. The excitations that would destroy any ordered phase in a thermal state are in fact localized in an MBL state and hence protecting order. This is famously now known as "localization protected quantum order" [67].

Discrete symmetries :

If the system has discrete symmetries, then the eigenstates can choose to break them spontaneously and become glassy at large disorder strengths [67–69]. For instance, in Ref [68], this was demonstrated in the disordered Ising spin chain whose Hamiltonian is given by,

$$H = -\sum_{i=1}^{L-1} J_i \hat{\sigma}_i^z \hat{\sigma}_{i+1}^z + h \sum_{i=1}^{L} \hat{\sigma}_i^x + J_2 \sum_{i=1}^{L-2} \hat{\sigma}_i^z \hat{\sigma}_{i+2}^z, \tag{3.10}$$

where $J_i = J + \delta J_i$ are independent positive random numbers with δJ_i drawn from uniform distribution $[-\delta J, \delta J]$. The model has a global $\mathbb{Z}_2$ symmetry implemented by the spin-flip operator $\hat{P} = \prod_i \hat{\sigma}_i^x$. Two distinct MBL phases can arise in this model. At weaker disorder, the model exhibits a trivial paramagnetic MBL phase. At large disorder, a spin glass phase can occur with long range order in an eigenstate $|\alpha\rangle$ with $\langle \hat{\sigma}_i^z \hat{\sigma}_j^z \rangle_\alpha \neq 0$ for $|i - j| \to \infty$. The phase diagram of this particular model has been numerically studied and a transition from thermal to MBL paramagnetic phase followed by a transition to an MBL spin-glass phase was seen on increasing the disorder strength. In the spin glass phase, the eigenstates spontaneously break the discrete symmetry and this manifests through the pairing of energy levels in the many-body spectrum of a finite sized system.

Interestingly, in Chapter 4, we numerically show in a model Hamiltonian with non-local interactions that the excited states display a transition into the spin glass phase while the ground state remains exceptional with no phase transition. We will argue that this is due to an emergent symmetry in the state. Likewise, new phases have also been studied in models with other discrete Abelian symmetries and also in higher dimensions.

Non-Abelian continuous symmetries:

It has been shown that the MBL phase is inconsistent with non-Abelian continuous symmetries [70, 11, 71, 72]. This is because of the multiplet structure demanded by the symmetry which causes degeneracies that ultimately lead to thermalization even at large disorder. A classic example where MBL breaks down is when the system has SU(2) symmetry. To understand this easily [71], consider two disconnected systems L and R. The SU(2) symmetry demands that each eigenstate $|\alpha\rangle_L$ of the system L has its own multiplet m_L with a total spin S_L with a degeneracy of $2S_L + 1$ and likewise for the other disconnected piece R. No matter how weakly one couples them together, the eigenstates of the full system (L+R) would transform as an irreducible representation of the SU(2) symmetry which acts on the combined system. Hence the eigenstates of the full system (L+R) would form larger multiplets with spins that can take values $|S_L - S_R|, |S_L - S_R| + 1, ..., S_L + S_R$ and utmost they can have an entanglement entropy $S \approx \ln(\min(S_R, S_L))$ and not an area law scaling [71].

In Chapter 5, we show the absence of MBL in a long ranged SU(2) invariant model even at large disorder which is in line with the above argument.

Emergent symmetries :

There can be an interesting scenario when certain eigenstates of a Hamiltonian have a particular symmetry, where as, the Hamiltonian does not enjoy that symmetry. These are called "emergent symmetries". The role of emergent non-Abelian continuous symmetries could be interesting in MBL systems that forbid MBL only in the eigenstates with symmetry and not preventing other eigenstates. We show in Chapter 4 where MBL is avoided in a single eigenstate of a disordered spin-1/2 chain that has an emergent SU(2) symmetry while not preventing the rest of the states from localizing.

3.2.2   Quantum many-body scars

In MBL, at strong disorder, all of the eigenstates in the many-body spectrum violate ETH and hence qualifies as an example of strong ergodicity breaking. However, recently, systems that violate ETH weakly, dubbed as " quantum many-body scars " have been much of interest with a plethora of different examples.

In a recent experiment in 2018, a family of Rydberg atom simulators revealed an unexpected dynamical phenomenon [73]. The experiment was essentially a set of two-level systems where an atom can be in either the ground state $|G\rangle$ or in the excited state $|E\rangle$. The system had a kinetic constraint that induced strong Van der Waals repulsion between atoms in the excited states leading to Rydberg blockade (excitations of neighbouring atoms in prohibited, eg., $|..EE...\rangle$) that was achievable by tuning the inter atomic distance. When specific initial states (such as density wave state, $|Z_2\rangle \equiv |EGEG...\rangle$) were chosen, surprising revivals were observed while the system showed quick relaxation otherwise. This strong dependence on the initial conditions was unexpected for the system which did not have any conserved quantities other than the total energy. Also, the model had no disorder and thus ruling out the possibility of integrability or localization. An effective Hamiltonian capturing the physics was constructed as below [74],

$$H = \sum_i \hat{P}_{i-1} \sigma_i^x \hat{P}_{i+1} \tag{3.11}$$

where $\sigma_i^x = |G\rangle_i\langle E|_i + |E\rangle_i\langle G|_i$ and $\hat{P}_i = |G\rangle_i\langle G|_i$ is the projector onto the ground state that enforces the Rydberg blockade and hence makes the model interacting. Numerical studies have observed energy level repulsion for the Hamiltonian indicating absence of integrability. Surprisingly, the spectrum of the Hamiltonian contained a measure zero set of eigenstates that had a significant overlap with the $|Z_2\rangle$ state. Furthermore, the entanglement entropy for these exact set of states were low despite living midst a sea of high energy states which had entropy close to the thermal value. It was these small number of ETH violating states that were responsible for the many-body revivals witnessed in the experiment. The special states were called "quantum many-body scars" and the phenomenon is analogous to the physics of single particle quantum scarring where the quantum wavefunction is enhanced along the unstable periodic trajectories for instance in the billiard stadium set up.

There are now several mechanisms responsible for quantum scarring in many-body systems and can be broadly categorized into three types [12]. Firstly, projector embedding method due to Shiraishi and Mori [75], allows target eigenstates $|\psi_i\rangle$ to be embedded at arbitrarily high energies. This method uses local projectors P_i that annihilate the target states $P_i|\psi_j\rangle = 0$ for any i ranging over lattice sites $1, 2, \cdots L$. Then one defines a Hamiltonian of the form $H = \sum_i^L P_i h_i P_i + H'$, where h_i acts near site i and $[H', P_i] = 0$ for all i. This Hamiltonian then has the target states at high energy density as the quantum scars and the other states in the spectrum obey ETH with a suitable choice of H'. Secondly, there is the method of spectrum generating algebra defined by a local operator $\hat{Q}^\dagger$ obeying the relation $([H, \hat{Q}^\dagger] - \omega\hat{Q}^\dagger)W = 0$, where ω is some energy scale and W is a linear subspace of the full Hilbert space which is invariant under the action of $\hat{Q}^\dagger$. Then, starting from an eigenstate $|\psi_0\rangle$ of H, the tower of scar eigenstates $(\hat{Q}^\dagger)^n|\psi_0\rangle$ with energy $E_0 + n\omega$ are built. As examples, quantum scars in the AKLT model [76] and other spin chains [77–80] were constructed using this method. Thirdly, there can be a situation when repeated action of the Hamiltonian H on some arbitrary vector in the Hilbert space which generates the sequence of Krylov vectors terminate after $n + 1$ steps. Hence, now if one looks at the dynamics initialized from any state within the Krylov subspace, at any later time, remains within the same subspace. This is known as "Krylov restricted thermalization" which leads to a dynamical fracture of the Hilbert space. Examples of this mechanism include fractional quantum Hall effect in a quasi one-dimensional limit [81] and in constrained hopping models of bosons in optical lattices [82, 83]. All these mechanisms bear a similarity in the emergence of a decoupled subspace within the many-body Hilbert space not related to any global symmetry of the full Hamiltonian.

The highly athermal nature of the scar states allow them to exhibit interesting properties such as long range order [77] and even topological order [84]. There has been a proposal to deform exactly solvable models such as the toric code so as to make the model non-integrable and with a topologically ordered scar state placed in the middle of the spectrum of ETH states [84]. In Chapter 6, we provide two other constructions of topologically ordered scarred models. In one case, we build Hamiltonians that realize scar states with chiral topological order by exploiting the existence of multiple operators derived from CFT that annihilate the scar state. In addition, we show that the scar states survive with small disorder and remain topologically ordered by showing the existence of anyons with well defined charges. In an other example, we describe a general strategy to turn classes of frustration free lattice models into similar classes containing topological quantum many-body scars within the bulk of their spectrum while preserving much or all of the original symmetry.

Chapter 4

Escaping many-body localization in a single eigenstate via an emergent symmetry

In Chapter 3, we saw that the interplay of disorder and interactions in many-body systems induces a localized phase where eigenstates violate ETH. At strong disorder, in 1D, all the eigenstates are localized. This is hence an example of strong violation of ETH or a fully many-body localized situation. One can ask if this is always the case or can one construct a model which exhibits a weak violation of MBL even with extreme disorder ? It turns out that this requires two ingredients, MBL and an emergent symmetry. Previously, we saw that a MBL phase is impossible with non-Abelian continuous symmetry groups. The essential idea would be to construct MBL Hamiltonians with emergent non-Abelian symmetries in eigenstates even in the presence of disorder. Upon adding large disorder, all but states with emergent non-Abelian symmetries localize.

In this chapter, we demonstrate this situation in a $\mathbb{Z}_2$ symmetric non-local model with an analytically known critical ground state with an emergent SU(2) symmetry. We show numerically that the states in the middle of the spectrum are glassy and many-body localized. We confirm this through computations of entanglement entropy and spin glass order parameter for states in the middle of the spectrum and the level spacing statistics. For the ground state alone, we show computations of Renyi entropy and full counting statistics for large system sizes, indicating that the state retains its Luttinger liquid behavior and remains non glassy even with significant disorder. By destroying the emergent symmetry, we show that the ground state becomes glassy and is no longer special. Finally, we also discuss how the ground state can be moved up in the spectrum portraying an instance of weak MBL violation

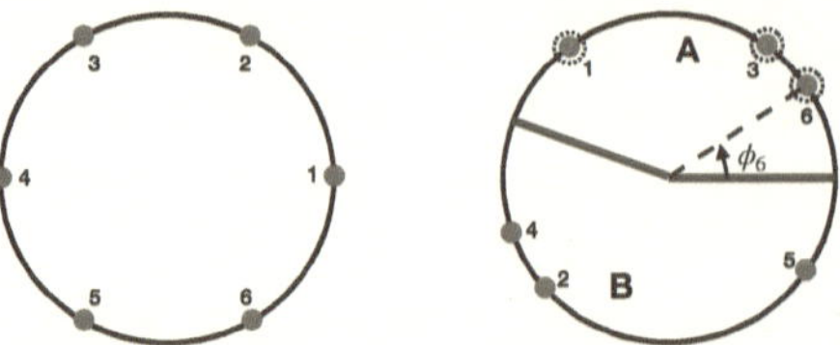

Fig. 4.1 Left: Uniform lattice on a circle with N=6 spins. Right: Disordered lattice with N=6 spins. When we divide the system into two parts to compute the entanglement entropy, we choose part A of the system to be the $N/2$ spins with the smallest angles ϕ_j and part B to be the remaining spins.

where a single non-MBL state is embedded in a spectrum of MBL states. This chapter is based on the following reference [85].

- " *Many-body delocalization via emergent symmetry* ", **N. S. Srivatsa**, R. Moessner & A. E. B. Nielsen, **Phys. Rev. Lett. 125, 240401 (2020)**

4.1   Model Hamiltonian

We study a system of interacting N spin-1/2 particles whose Hamiltonian is given by,

$$H = \sum_{i \neq j} F_{ij}^A (S_i^x S_j^x + S_i^y S_j^y) + \sum_{i \neq j} F_{ij}^B S_i^z S_j^z + F^C \tag{4.1}$$

In the above expression, $S_i^a = \sigma_i^a/2$, $a \in \{x,y,z\}$, where σ_i^a are the Pauli matrices acting on the ith spin. The coupling strengths are given by,

$$\begin{aligned}
F_{ij}^A &= -2w_{ij}^2, \\
F_{ij}^B &= -2w_{ij}^2 + 2w_{ij}\left(\sum_{k(\neq i)} w_{ik} - \sum_{k(\neq j)} w_{jk}\right), \\
F^C &= \frac{N(N-2)}{2} - \frac{1}{2}\sum_{i \neq j} w_{ij}^2,
\end{aligned} \tag{4.2}$$

where, $w_{jk} = -i/\tan[(\phi_j - \phi_k)/2]$ and the positions of the spins are constrained to live on the unit circle given by $e^{i\phi_j}$. We introduce disorder in the model by randomly positioning the spins on the unit circle by choosing,

$$\phi_{f(j)} = 2\pi(j + \alpha_j)/N, \quad j = 1,2,...,N, \tag{4.3}$$

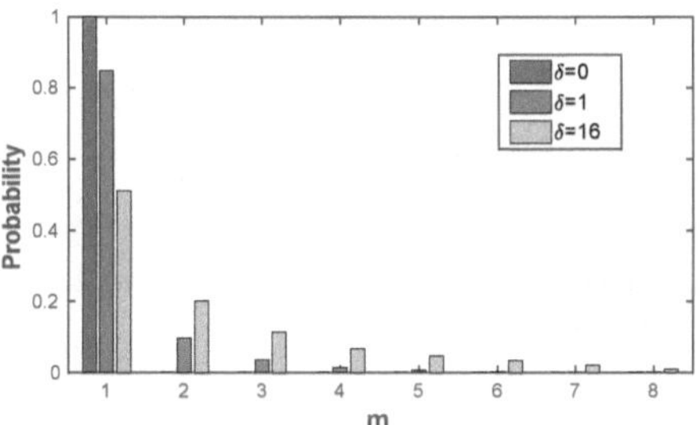

Fig. 4.2 Probability that the spin i interacts most strongly with one of the spins $i \pm m$ modulus N for different disorder strengths. This is computed for $N = 16$ spins.

where α_j is a random number chosen with constant probability density in the interval $[-\frac{\delta}{2}, \frac{\delta}{2}]$ and $\delta \in [0, N]$ is the disorder strength. We choose the indices $f(j) \in 1, 2, ..., N$ such that the spins are always numbered in ascending order when going around the circle. For the clean case, $\delta = 0$, the spins are uniformly distributed on the circle, and for maximal disorder, $\delta = N$, all the spins may be anywhere on the circle as depicted in Fig. 4.1. For $\delta \geq 1$, neighboring spins can be arbitrarily close. The Hamiltonian (7.10) is fully connected, meaning that every spin interacts with every other spin. If we interpret a spin up as a particle and a spin down as an empty site, the first term in the Hamiltonian is a hopping term while the second term is the interaction between spins. For short distances (when $|\phi_i - \phi_j|$ is small), the coefficient F^A decays quickly with increasing $|\phi_i - \phi_j|$ as $F^A_{ij} \propto \frac{1}{|\phi_i - \phi_j|^2}$. The behavior of the coefficient F^B_{ij} of the two body interaction is complicated. The maximum interaction strength for spin i does not need to be with one of the spins $i \pm 1$ modulus N. Hence, we compute the probability that the spin i interacts most strongly with one of the spins $i \pm m$ modulus N which is shown in Fig. 4.2. For the case with disorder ($\delta \neq 0$), the Hamiltonian has no SU(2) symmetry and has only a global discrete $\mathbb{Z}_2$ splin flip symmetry generated by the operator $\prod_{i=1}^{N} \sigma_i^x$.

4.2   Physics of the ground state with disorder

In the zero magnetization sector ($\sum_i S_i^z = 0$), the Hamiltonian in Eq. (7.10) can be written as,

$$H = -2 \sum_k \Lambda_k^\dagger S_k^z \Lambda_k \tag{4.4}$$

where, $\Lambda_k = \sum_{j(\neq k)} w_{kj} [S_j^x - iS_j^y - 2(S_k^x - iS_k^y)S_j^z]$ is an operator that annihilates the state,

Fig. 4.3 The ground state of a random Heisenberg model which consists of singlets connecting spins over arbitrarily long length scales.

$$|\psi_0\rangle \propto \sum_{s_1,\ldots,s_N} \delta_s \prod_{i<j} \sin[(\phi_i - \phi_j)/2]^{(s_i s_j - 1)/2} \prod_k e^{i\pi(k-1)(s_k+1)/2}|s_1,\ldots,s_N\rangle, \tag{4.5}$$

where, $|s_1,\ldots,s_N\rangle$ is a many-body state of spins with $s_i \in \{-1,1\}$ and $\delta_s = 1$ only if $\sum_i s_i = 0$ and $\delta_s = 0$ otherwise. Hence, $|\psi_0\rangle$ is the exact zero energy state of the above Hamiltonian. Furthermore, we numerically find that the state in Eq. (4.5) is the non-degenerate ground state in the zero magnetization sector and at zero disorder ($\delta = 0$), this corresponds to the ground state of the famous Haldane-Shastry (HS) model defined in Eq. (2.16). We will refer to the state with $\delta \neq 0$ as the disordered Haldane-Shastry (HS) state which is a singlet of the total spin and hence SU(2) invariant while the Hamiltonian is not, i.e,

$$\begin{aligned} \mathbf{S}_{\text{tot}}|\psi_0\rangle &= 0 \\ [H, \mathbf{S}_{\text{tot}}] &\neq 0. \end{aligned} \tag{4.6}$$

Hence, the SU(2) symmetry in the ground state is *emergent* which will lead to a special behavior only for the ground state in the model with disorder.

As we saw in the previous chapter, the low energy physics of the Haldane-Shastry model is described by the Luttinger liquid theory. In 1D, random spin chains with short range interactions such as the random spin-1/2 Heisenberg chain, are known to flow to a random singlet phase (RSP) with quantum critical properties. In this phase, the ground state is best described by a configuration where pairs of strongly interacting spins form singlets, which can span arbitrarily large distances as depicted in Fig. 4.3. This is a localized phase with typical spin-spin correlations decaying as $C^{\text{typ}}(r) \sim e^{-c\sqrt{r}}$. The RSP is essentially an infinite randomness fixed point which is a random version of pure CFT which show universal behavior in entanglement entropy.

The question we want to address is the fate of the disordered HS state. The possibilities are, either the state can loose its quantum critical properties and become many-body localized

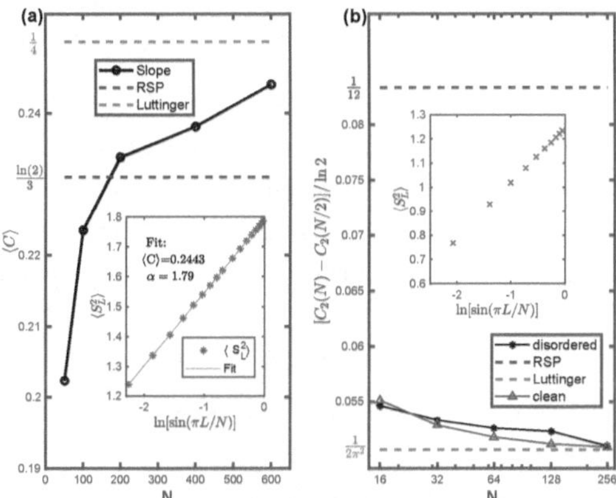

Fig. 4.4 (a) The main plot shows the disorder averaged coefficient $\langle C \rangle$ as a function of the system size N for $\delta = 0.75$. The result might suggest that the state may be a Luttinger liquid, but the method is not accurate enough to make clear conclusions. We use 10^3 disorder realizations, and the error from the Monte Carlo simulations and disorder averaging is of the order 10^{-4}. Inset shows the plot of the Renyi entropy $S_L^{(2)}$ for $N = 600$ and $\delta = 0.75$ computed for different subsystem sizes L which follows the logarithmic relation given in Eq. (4.9). (b) The coefficient ξ in the second cumulant C_2 computed for different systems divided into two halves is close to the value for the Luttinger liquid both for the clean ($\delta = 0$) and the disordered ($\delta = 4$) Haldane-Shastry state. Each data point is averaged over 10^5 disorder realizations. The inset shows the logarithmic scaling of the averaged Renyi entropy for $\delta = 4$ and $N = 50$.

or retain its quantum critical behavior by either stablilizing the Luttinger liquid behavior or flow to a RSP.

4.2.1   Renyi entropy

We first discuss the computations for Renyi entropy performed on the HS state with disorder. The Renyi entropy of order 2 is defined as,

$$S_L^{(2)} = -\ln \operatorname{Tr} \rho_L^2, \tag{4.7}$$

where, ρ_L is the reduced density matrix of subsystem of size L. Since we know the analytical form of the ground state, we can perform Renyi entropy computations for considerably large system sizes invoking the replica trick by rewriting,

$$e^{-S_L^{(2)}} = \sum_{n,n',m,m'} |\langle n,m|\psi_0\rangle|^2 |\langle n',m'|\psi_0\rangle|^2 \frac{\langle\psi|n',m\rangle\langle\psi|n,m'\rangle}{\langle\psi|n,m\rangle\langle\psi|n',m'\rangle} \tag{4.8}$$

were $|n\rangle$ is an orthonormal basis for the L spins while $|m\rangle$ is an orthonormal basis for the rest $N-L$ spins. Now, this expression can be computed numerically for two independent spin chains and thereby carrying out the sum using the Monte Carlo techniques. For a quantum critical system, it is known that the Renyi entropy shows a universal behavior given by,

$$S_L^2 = C \ln[\sin(\frac{\pi L}{N})] + \alpha. \tag{4.9}$$

In the above expression, C is a universal constant that takes the value $1/4$ for the Luttinger liquid and $\ln(2)/3$ for the random singlet phase respectively and α is a non-universal constant. Monte Carlo simulations for $\delta = 0.1$, $\delta = 0.5$, and $\delta = 0.75$ in Ref. [86] showed that C might be closer to $\ln(2)/3$ than $1/4$ for $\delta = 0.5$ and $\delta = 0.75$ indicating that the state might flow towards the RSP. However, in their computations, only L values close to $N/2$ were used for the fitting to extract the value C. We redo the computations for $\delta = 0.75$ and extract the disorder averaged value of the coefficient C by performing a fit by choosing L values in the entire range (1 to $L/2$) for different system sizes N. Our results suggest that C might go to $1/4$ for large system sizes as shown in Fig. 4.4 (a). However, the main conclusion from the computations is that the uncertainty in determining C due to, e.g., finite size effects and ambiguity in the fitting procedure is not small compared to the difference between $1/4$ and $\ln(2)/3$. Hence we need to further check this with other probes.

4.2.2 Full counting statistics

Another useful probe to understand the quantum critical properties of a quantum many-body system is to study the fluctuations of macroscopic observables through the full distribution of charges known as the full counting statistics (FCS). For a charge operator O, the FCS is defined as,

$$\chi(\lambda) = \langle e^{i\lambda O}\rangle = \sum_{t=0}^{\infty} \frac{(i\lambda)^t}{t!} \langle O^t\rangle. \tag{4.10}$$

However, the physically relevant quantity is the second cumulant of the charge operator given by, $C_2(N/2) = \langle O^2\rangle - \langle O\rangle^2$. FCS can also be used as a probe to determine the universality class of the ground states of spin chains by studying the fluctuations of the total magnetization of half of the subsystem, $M = \sum_{i=1}^{N/2} S_i^z$. In particular, for a quantum critical system, the second

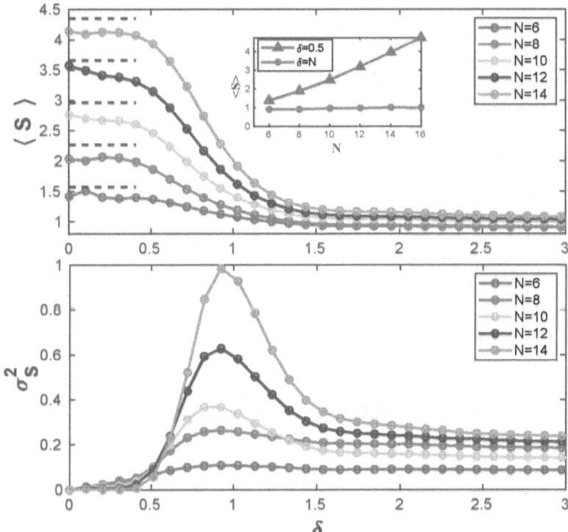

Fig. 4.5 Top: The transition to the MBL phase is seen in the entanglement entropy S of half of the chain for the state closest to the middle of the spectrum averaged over 10^5 disorder realizations as a function of the disorder strength δ for different system sizes N. The blue dashed lines indicate the thermal value $[N \ln(2) - 1]/2$ of the entropy. The inset shows that the mean entropy follows a volume law for weak disorder and an area law for strong disorder. Bottom: The variance σ^2 of the entanglement entropy computed from the same set of data shows a peak at the transition point.

cumulant of M is known to diverge logarithmically with system size given by,

$$C_2(N/2) \sim \xi \ln(N/2) + \beta \tag{4.11}$$

where, ξ is a universal quantity that takes a value $1/(2\pi^2)$ for a Luttinger liquid and $1/12$ for a RSP and β is a non universal constant. To compute $C_2(N/2)$, we make use of the following set of linear equations derived in Ref. [87] to compute the two point correlations.

$$w_{ij}\langle S_i^z S_j^z \rangle + \sum_{k(\neq i,j)} w_{ik}\langle S_j^z S_k^z \rangle + \frac{3}{4}w_{ij} = 0 \quad i \neq j \tag{4.12}$$

We are thus able to numerically extract the coefficient ξ for quite large system sizes and we find that even for a large disorder strength $\delta = 4$, the coefficient ξ converges to $1/2\pi^2$ for large system size as shown in Fig. 4.4 (b). This confirms that the disordered HS state indeed stabilizes the Luttinger liquid phase even for large disorder strengths.

We are now hence sure that the ground state is quantum critical and does not many-body localize even for large disorder strengths. This is in agreement with the violation of area law scaling of entanglement entropy in a single eigenstate due to SU(2) symmetry.

4.3   Physics of the excited states

Next, we probe whether the excited states become many-body localized or not. We recall that the Hamiltonian we have is non-local with hopping terms that decay quickly but the interactions do not. The existence of MBL in systems with long ranged terms is still an interesting open question. Perturbative arguments suggest that the MBL phase is destroyed with the presence of hopping terms that do not decay faster that

4.3.1   Entanglement entropy

A standard probe to check for MBL is entanglement. As discussed in previous chapters, MBL eigenstates have very low entanglement, typically area-law entanglement. We compute the standard Von Neumann entropy defined as,

$$S = -\text{Tr}[\rho \ln(\rho)], \tag{4.13}$$

for an eigenstate $|\psi\rangle$. In the above expression, $\rho = \text{Tr}_B(|\psi\rangle\langle\psi|)$ is the reduced density operator after tracing out part B of the chain as shown in Fig.4.1. For each disorder realization, we pick the eigenstate whose energy is closest to the middle of the spectrum, i.e., closest to

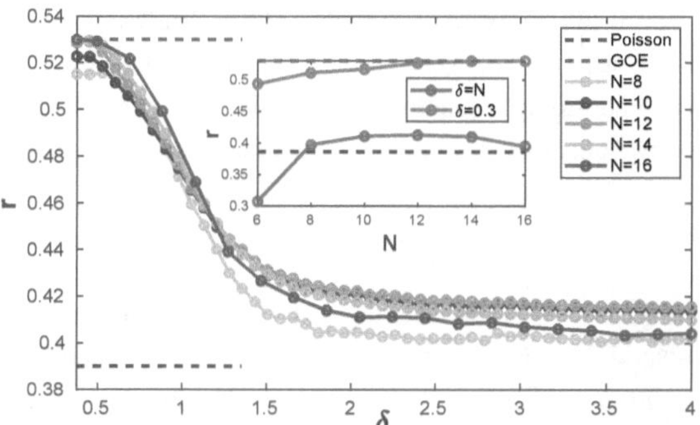

Fig. 4.6 The adjacent gap ratio r shown for different disorder strengths and system sizes N. The r-value is close to the GOE for weak disorder and close to Poisson distribution for large disorder indicating MBL. The disorder averaging is performed over 10^3 samples for $N = 16$ and 10^4 samples for other system sizes. The inset shows the disorder averaged r-value computed for large and small disorder strengths for different system sizes. For small disorder ($\delta = 0.3$), a convergence to 0.53 is seen indicating an ergodic phase and at maximum disorder ($\delta = N$), convergence to 0.39 is seen indicating an MBL phase.

the energy $(E_{\max} + E_{\min})/2$, where $E_{\min}$ is the ground-state energy and $E_{\max}$ is the highest energy in the spectrum. In Fig. 4.5, we plot the mean entanglement entropy and the variance of the distribution as a function of the disorder strength δ for different system sizes. The plots show a phase transition form an ergodic phase to a non-ergodic MBL phase at $\delta \approx 1$. The reason why the phase transition happens around that point could be due to the fact that for $\delta \geq 1$, neighboring spins can be arbitrarily close, while this is not the case for $\delta < 1$. The inset in the Fig. 4.5 shows how the disorder averaged entropy $\langle S \rangle$ scales with the system size N on the weak and strong disorder side of the transition. At small disorder, the entropy grows with the system size, and the slope is comparable to the slope for the entropy of an ergodic system. At large disorder, the entropy remains roughly a constant with system size indicating area-law scaling of entanglement. This indicates a non-ergodic MBL phase at large disorder strengths.

4.3.2   Level-spacing statistics

The level spacing statistics is another helpful tool to identify whether a system is in an ergodic phase, a many-body localized phase, or in some other phase. Hamiltonians with or without time reversal symmetry in the ergodic phase are known to exhibit Gaussian orthogonal ensemble (GOE) or Gaussian unitary ensemble (GUE) level spacing statistics, respectively, indicating the presence of energy level repulsion [88]. While in the localised phase, the level spacing statistics obey the Poisson distribution indicating the absence of level repulsion. One can compute a dimensionless quantity, which is the ratio of consecutive gaps of distinct energy levels [89]. The quantity of interest is called the adjacent gap ratio and is defined as follows

$$r = \frac{1}{N_s} \sum_n \frac{\min(\delta_n, \delta_{n-1})}{\max(\delta_n, \delta_{n-1})}, \tag{4.14}$$

where $\delta_n = E_{n+1} - E_n$ and N_s is the total number of states in the spectrum. The value of r averaged over several disorder realizations is around 0.529 for the Gaussian orthogonal ensemble (GOE) and around 0.386 for the Poisson distribution. We recall that we are considering the sector of the Hilbert space with zero net magnetization. Also, since the Hamiltonian commutes with the parity operator $\prod_i \sigma_x^i$, we consider the energies in one of the parity sectors for computing the level spacing statistics. The Fig. 4.6 shows the disorder averaged value of the adjacent gap ratio r for various disorder strengths and system sizes. The inset in the Fig. 4.6 shows that the r value is close to 0.53 for small disorder and close to 0.39 for large disorder and for the biggest system size studied. This indicates that the system is indeed in an MBL phase at large disorder strengths.

4.3.3 Glassiness

The fact that pairs of almost degenerate states with opposite parity appear in the spectrum suggests that there is spin glass order in the excited states. In an eigenstate $|\psi_n\rangle$ this can be identified by the divergence of an Edwards-Anderson order parameter given by,

$$\chi^{\text{SG}} = \frac{1}{N} \sum_{i \neq j}^{N} \langle \psi_n | \sigma_i^z \sigma_j^z | \psi_n \rangle^2. \tag{4.15}$$

For eigenstates in the middle of the spectrum, we find (see Fig. 4.7(a)) that there is glassiness for strong disorder ($\langle \chi^{\text{SG}} \rangle$ increases with system size), but not for weak disorder ($\langle \chi^{\text{SG}} \rangle$ approaches zero with increasing system size). We perform a finite size scaling analysis to get an estimate of the critical disorder strength. The scaling parameters are given in Fig. 4.7(b), where we define the scaling function as $x_L = (\delta - \delta_c)N^{\frac{1}{\nu}}$ and $y_L = \langle \chi^{\text{SG}} \rangle / N^a$. Glassiness sets in at around the same disorder strength ($\delta_c \approx 1$) as MBL. Remarkably, the ground state stands out special and the spin glass order parameter takes the value $\chi^{\text{SG}} = 1$ for all disorder strengths δ and system sizes N. We show this analytically for $N = 2, 4, 6$ and numerically for higher N in the Appendix B.

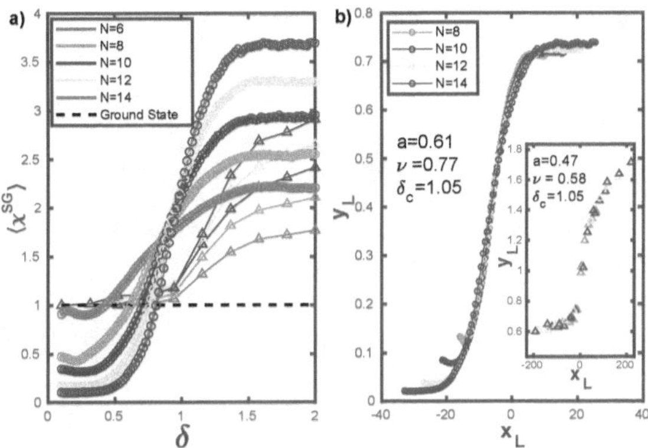

Fig. 4.7 (a) Disorder averaged spin glass order parameter $\langle \chi^{SG} \rangle$ for a state in the middle of the spectrum ($E/E_{\max} = 0.5$, circles), a low energy state ($E/E_{\max} = 0.01$, triangles), and the ground state ($E = 0$, dashed line) as a function of the disorder strength δ. The excited states show glassiness for strong disorder, and the ground state is not glassy. (b) A finite size scaling collapse for the state in the middle of the spectrum. The inset shows the scaling collapse for the low energy state. For both cases, the predicted phase transition point is $\delta_c \approx 1.05$

4.4   Emergent symmetry and weak violation of MBL

The ground state of the model is clearly special because of the emergent SU(2) symmetry that prohibits MBL and glassiness. The question thus arises if it is only the ground state that is special or to some extent states close to the ground state also inherit similar behavior. We computed the disorder averaged spin glass order parameter for all the states in the spectrum at large and small disorder as shown in Fig. 4.8. For small disorder all states show low values for the spin glass order parameter. For large disorder, states in the middle of the spectrum appear glassy but few low lying states seem to have a different behavior. For the system size studied, it looks likely that there are few states that behave similar to the ground state but this warrants further studying large systems. We also show that the special behavior of the ground state disappears together with the emergent SU(2) symmetry. To do so, we slightly modify the hopping strengths F_{ij}^{A} to $-1.9w_{ij}^{2}$. This preserves the Z_2 symmetry of the Hamiltonian, but not the emergent SU(2) symmetry of the ground state. Fig. 4.8(b) shows that the ground state is now glassy for strong disorder. If instead we add a small amount of the Haldane-Shastry Hamiltonian, the ground state is unaltered, and the Z_2 symmetry of the Hamiltonian is preserved. In this case, the spin glass order parameter behaves similarly to the results in Fig. 4.7 (a). One can actually engineer a situation where the ground state of

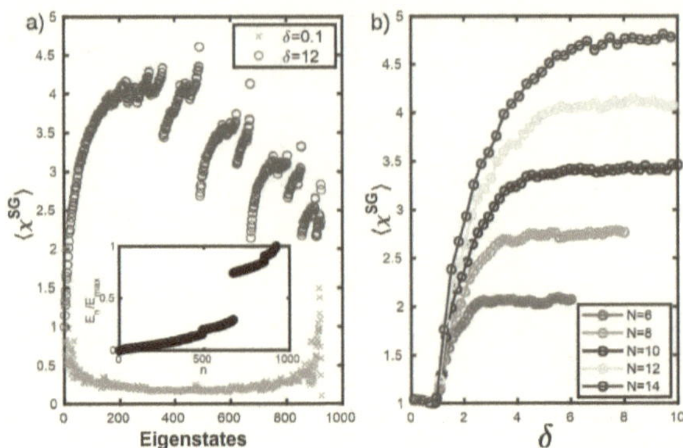

Fig. 4.8 a) $\langle \chi^{SG} \rangle$ for all eigenstates for a system with 12 spins and weak ($\delta = 0.1$) or strong ($\delta = 12$) disorder. The states close to the ground state have different values of $\langle \chi^{SG} \rangle$ compared to the states in the middle of the spectrum. The inset shows that the disorder averaged energy spectrum (normalised) at $\delta = 12$ has gaps, and these coincide with the jumps in $\langle \chi^{SG} \rangle$. b) When we destroy the emergent SU(2) symmetry by perturbing the Hamiltonian (we modify F_{ij}^A to $-1.9w_{ij}^2$), the ground state becomes glassy for $\delta \geq 1$. In all cases, the number of disorder realizations is 10^4.

H lies somewhere in the middle of the spectrum of MBL states by studying the following Hamiltonian,

$$\tilde{H} = (H - \beta)^2, \tag{4.16}$$

where $\beta \in]0, E_{\max}/2[$ can be chosen to place the state anywhere in the spectrum as shown in Fig. 4.9. This is now a situation where a non glassy, quantum critical state lives in the middle of the spectrum of localized glassy states and is a weak violation of MBL. If the delocalized state should turn out to be genuinely isolated in a sea of localized ones, one could compare this behavior to the celebrated many-body scars. In the former case, a state with above-area-law entanglement is embedded in a sea of area law entangled states, while in the latter, one or several nonergodic states are embedded in a sea of ergodic states.

4.5   Conclusion

In this chapter, we have constructed a rather different kind of MBL model where the ground state, due to an emergent symmetry, is protected from MBL. We showed that the states at a finite energy density display a transition from an ergodic to a MBL glass phase while

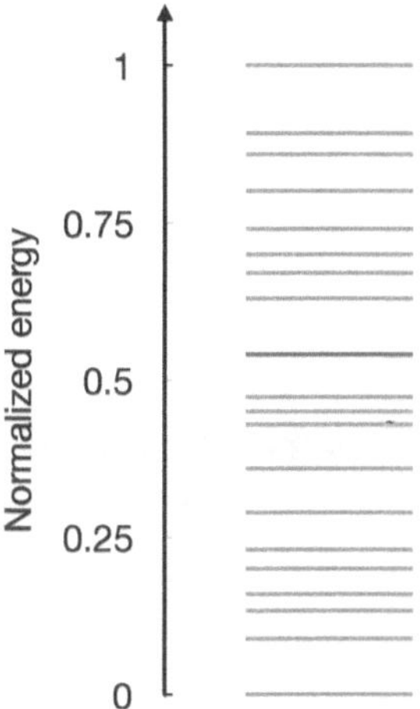

Fig. 4.9 The normalized energy spectrum of the Hamiltonian $\tilde{H}$ with $N = 6$, $\delta = 2$, $\beta = 0.4$. The non-glassy quantum critical state is shown red in color while MBL glassy states are shown blue in color. This is hence a scenario of weak MBL violation.

the ground state remains quantum critical with Luttinger liquid properties. We tested the role of emergent symmetry by demonstrating the ground state becoming glassy when the emergent symmetry is destroyed by a small perturbation. We also saw that the model has the unexpected property that the disorder averaged energy spectrum has gaps. The background for this is not understood and would be interesting to investigate further. Finally, we discussed how the ground state may be moved up to the middle of the spectrum which would now give a flavour of an inverted quantum scar and hence a weakly broken MBL phase.

Chapter 5

The inhomogeneous Haldane-Shastry model

In the previous chapter, we investigated an interesting scenario where MBL was avoided in a single eigenstate due to an emergent SU(2) symmetry. As already discussed in Chapter 3, SU(2) symmetry in a model forbids many-body localization even in the presence of large disorder [11]. An investigation of the antiferromagnetic Heisenberg chain with SU(2) symmetry and nearest neighbor exchange interactions of random strengths showed that a different type of non-ergodic phase appears in a broad regime at strong disorder [71, 72]. In this chapter, we are interested in the question, whether a non-ergodic, low entanglement entropic phase can also appear in models with non-local terms.

In this chapter, we study a generalization of the Haldane-Shastry model that was proposed in Refs. [86, 87]. In the generalized version, the spins can be anywhere on the unit circle and the ground state is the disordered Haldane-Shastry (HS) state described in Eq. 4.5. We investigate the effects of disorder in this model by randomizing the position of the spins on the unit circle. At large disorder, we see that the interactions do not have a simple power law decay. This is an interesting situation with an interplay of SU(2) symmetry, long ranged interactions and disorder. We show that the model avoids being MBL even at large disorder due to the SU(2) symmetry, but displays an emergent non-ergodic phase despite having non-local interactions between the spins. We confirm this non-ergodic phase through the computations of level spacing statistics and mid spectrum entanglement entropy. This chapter is based on the following reference [90].

- " *Disordered Haldane-Shastry model* ", S. Pai*, **N. S. Srivatsa***[†] & A. E. B. Nielsen, **Phys. Rev. B 102, 035117 (2020)** [* authors contributed equally to this work]

5.1 Model and disorder

We first recall the generalized Haldane-Shastry model from [87, 86] given by,

$$H = -2 \sum_{i \neq j} \left[\frac{z_i z_j}{(z_i - z_j)^2} + \frac{(z_i + z_j)(c_i - c_j)}{12(z_i - z_j)} \right] \vec{S}_i \cdot \vec{S}_j, \tag{5.1}$$

The above Hamiltonian describes a system of N spin-1/2 particles sitting on the unit circle as illustrated in Fig. 2.1. In the original HS model, $z_j = e^{i2\pi j/N}$, which means that the spins are uniformly spaced. We shall here consider the more general case, where $z_j = e^{i\phi_j}$ with $\phi_j \in [0, 2\pi[$. In this expression,

$$c_j = \sum_{k \in \{1,2,\dots,N\} \setminus \{j\}} \frac{z_k + z_j}{z_k - z_j} \tag{5.2}$$

and $\vec{S}_i$ is the spin 1/2 operator (S_i^x, S_i^y, S_i^z) defined at the ith site. We write the Hamiltonian in (5.1) as

$$H = \sum_{i \neq j} h_{ij} \vec{S}_i \cdot \vec{S}_j. \tag{5.3}$$

The Hamiltonian is SU(2) invariant and conserves the net magnetization. We shall assume throughout that the net magnetization is fixed to zero. We add disorder to this Hamiltonian by defining the lattice positions z_j as,

$$z_j = e^{2\pi i (j + \alpha_j)/N}, \qquad j \in \{1, 2, \dots, N\}. \tag{5.4}$$

Here, α_j is a random number chosen with constant probability density in the interval $[-\frac{\delta}{2}, \frac{\delta}{2}]$, and $\delta \in [0, N]$ is the disorder strength. An example for $N = 6$ spins is shown in the right panel of Fig. 2.1. For disorder strength $\delta = 0$, the uniform model is recovered. When $0 < \delta < 1$, there is some disorder, but the spins are still ordered from 1 to N on the circle, and there is still one spin per $2\pi/N$ length of the circle. If we increase the disorder strength even further to $\delta > 1$, the order of the spins can change. For the maximal disorder strength $\delta = N$, all of the spins can be anywhere on the circle. It is seen that disordering the positions of the spins leads to a disordering of the coupling strengths. The coupling strengths can be arbitrarily large when the spins come close to each other. Fig. 5.1 shows the behavior of the coupling coefficients with distance for cases with and without disorder. Although the coupling coefficients quickly decay to zero for the clean case, this is no longer true at strong disorder as seen in the figure.

The ground state of this Hamiltonian in the zero magnetization sector coincides with the ground state of the famous HS model when the spins are positioned evenly on the unit

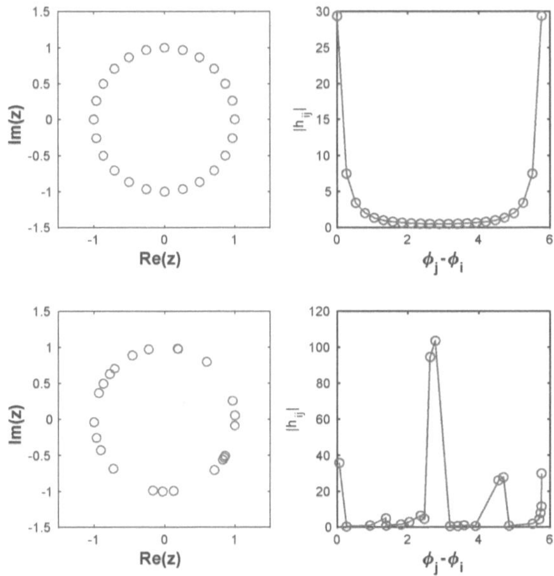

Fig. 5.1 Top left: Uniform lattice with $N = 20$ spins. Top right: Behavior of the coupling strengths $|h_{ij}|$ in the Hamiltonian on the clean lattice. Bottom left: One realization of the disordered lattice with $N = 20$ spins for maximum disorder strength $\delta = N$. Bottom right: Behavior of the coupling strengths for the disordered lattice shown on the left. In both the clean and disordered cases, we fix the ith spin to be the one with the smallest angle ϕ_i and plot $|h_{ij}|$ as a function of $\phi_j - \phi_i$. Note that for the disordered case, the coupling strength can be quite large between the spins that are far from each other.

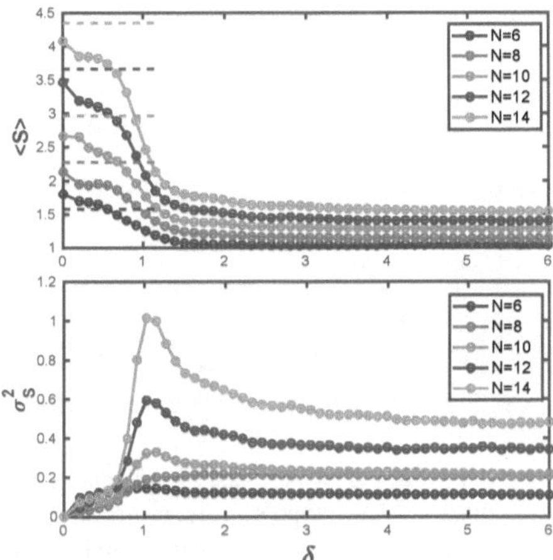

Fig. 5.2 Top: Entanglement entropy S of half of the chain for the state closest to the middle of the spectrum of the Hamiltonian (5.1) averaged over 10^4 disorder realizations as a function of the disorder strength δ. The different curves are for different system sizes N as indicated. The entropy $[N\ln(2) - 1]/2$ for an ergodic system is shown with horizontal dashed lines. Bottom: Variance σ^2 of the entanglement entropy computed from the same set of data. It is seen that the transition happens around $\delta = 1$.

circle and for the general case, matches the ground state of the non-SU(2) invariant model discussed in the previous chapter given by Eq. 4.5.

5.2 Properties of the disordered model

In the following, we study the properties of the highly excited states. The Hamiltonian can be truly long-ranged at strong disorder, and we find that the system is non-ergodic, but not many-body localized. This is similar to the findings in short-ranged SU(2) invariant models discussed in [71, 72]. In particular, we investigate numerically the entanglement entropy and the level spacing statistics.

5.2.1 Entanglement entropy

We compute the standard Von Neumann entropy defined in Eq. (4.13) for states in the middle of the spectrum. The way we partition the spin chain to compute the reduced density matrix of a subsystem is discussed in Chapter 4. In Fig. 5.2, we plot the mean entanglement entropy

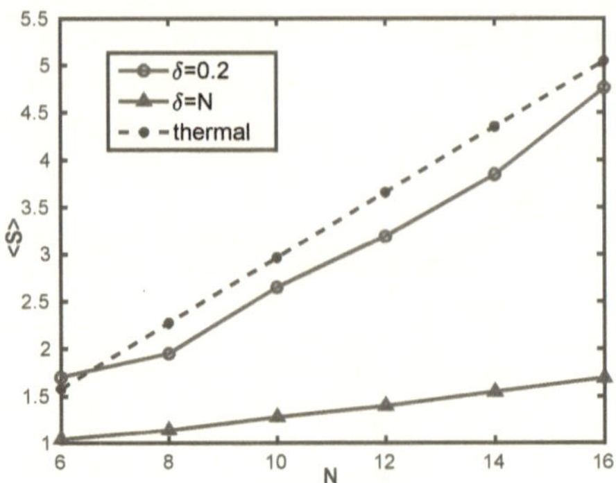

Fig. 5.3 Scaling of the disorder averaged half chain entanglement entropy $\langle S \rangle$ with the system size N for weak and strong disorder. The slope of a linear fit for the points computed at weak disorder is around 0.3, while for those computed at maximum disorder it is around 0.06. The dashed blue line shows the entropy $[N\ln(2) - 1]/2$ of an ergodic system for comparison. The results shown are averaged over 10^4 disorder realizations.

and the variance of the distribution as a function of the disorder strength δ for different system sizes. The plots show a phase transition at $\delta \approx 1$. The reason why the phase transition happens around that point could be due to the fact that for $\delta \geq 1$, neighboring spins can be arbitrarily close, while this is not the case for $\delta < 1$.

The Figure 5.3 shows how $\langle S \rangle$ scales with the system size N on the weak and strong disorder side of the transition and compares these results to the entropy of an ergodic system of N spin-1/2, which was conjectured to be $[N\ln(2) - 1]/2$ for $2^{N/2} \gg 1$ in [57]. At small disorder, the entropy grows with the system size, and the slope is comparable to the slope for the entropy of an ergodic system. At large disorder, the entropy again increases linearly with the system size, which shows that the system is not many-body localized. The slope is, however, much less than it is for weak disorder, and the system is therefore also not ergodic.

5.2.2 Level spacing statistics

The energy level spacing distribution of the clean Haldane-Shastry model has been studied in the past and is found not to obey the Poissonian distribution although the model is completely integrable [91, 92]. We compute the distribution $P(E_{n+1} - E_n)$ of the energy level spacings $E_{n+1} - E_n$ of the disordered Haldane-Shastry model for the maximally disordered case and find that, interestingly, it is neither of the Wigner-Dyson, nor of the Poisson kind, which is seen in Fig. 5.4.

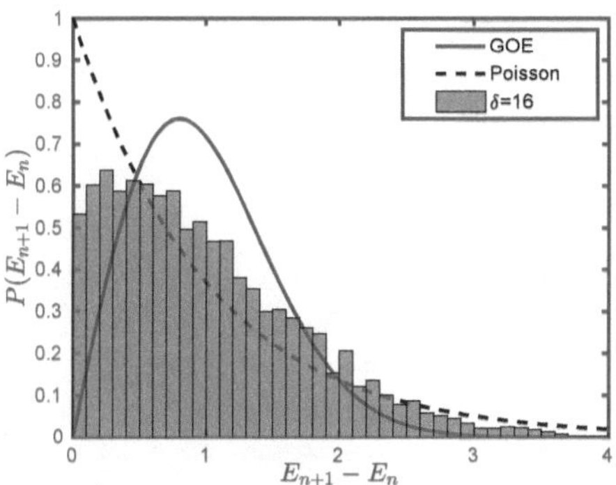

Fig. 5.4 The energy level spacing distribution of the disordered Haldane-Shastry model computed for 16 spins for one disorder realization at maximal disorder strength. It follows neither the Poisson distribution, nor the Gaussian orthogonal ensemble.

We also compute the adjacent gap ratio defined in Eq. (4.14). We recall that we are considering the sector of the Hilbert space with zero net magnetization. Also, since the Hamiltonian commutes with the parity operator $\prod_i \sigma_x^i$, we consider the energies in one of the parity sectors for computing the level spacing statistics. The plot in Fig. 5.5 shows the gap ratio computed for several system sizes at maximum disorder strength. The gap ratio is not close to either 0.529 or 0.386. This indicates that the system is in a non-ergodic phase that does not exhibit many-body localization. This is consistent with the sub-thermal value of the half-chain entanglement entropy at strong disorder observed above.

We avoid showing the level spacing statistics for weak disorder for the following reason. The clean Haldane-Shastry model has an enhanced Yangian symmetry generated by the total spin operator and the rapidity operator that both commute with the Hamiltonian [49], and this leads to degeneracies in the spectrum. For weak disorder, the model has no exact Yangian symmetry, but there are still approximate degeneracies visible in the energy spectrum. This is seen in Fig. 5.6, where we compare the energy spectra of the clean model and the disordered model at weak disorder. The energy spectrum with little disorder closely resembles the spectrum of the clean Haldane-Shastry model. The degeneracies, although not exact, have an impact on the level spacing statistics and make this measure more difficult to interpret.

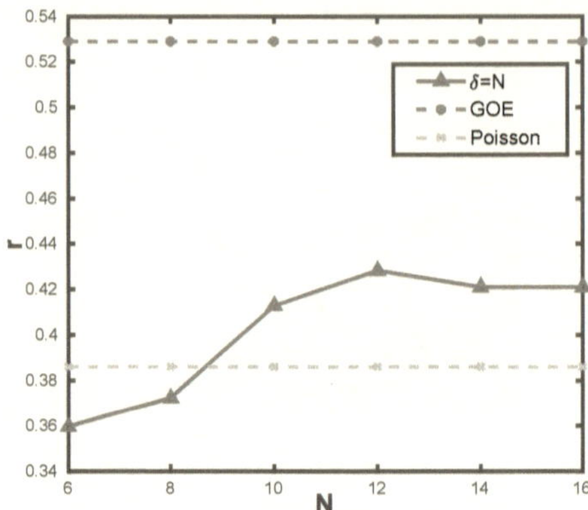

Fig. 5.5 The adjacent gap ratio r for the spectrum of the Hamiltonian (5.1) as a function of the system size N at maximal disorder strength. The results shown are averaged over 10^4 disorder realizations. The curve neither approaches the Poisson value, nor the value for the Gaussian orthogonal ensemble.

5.3   Adding disorder to the excited states of the Haldane-Shastry model

All the excited states of the clean model can be analytically expressed in terms of the positions z_j [93]. This opens up the possibility to add disorder into the excited states by modifying z_j. It turns out, however, that the resulting states $|\psi_n'\rangle$ are not eigenstates of the Hamiltonian (5.1) as soon as $\delta > 0$. If the states had been orthonormal, one could construct an alternative Hamiltonian as $H = \sum_n E_n |\psi_n'\rangle\langle\psi_n'|$, where E_n is the energy of the state $|\psi_n'\rangle$. We find numerically, however, that most of the states are not orthogonal for $\delta > 0$. The most appropriate way to add disorder into the model is hence to introduce it in the Hamiltonian (5.1) and study the properties of the eigenstates numerically as discussed in the previous sections.

5.4   Conclusion

In this chapter, we investigated the role of disorder on the Haldane-Shastry model, which has long-range interactions and SU(2) symmetry. The disorder averaged entanglement entropy for a state close to the middle of the energy spectrum showed a transition into a non-ergodic phase at strong disorder. The non-ergodic phase is not a many-body localized phase, since it

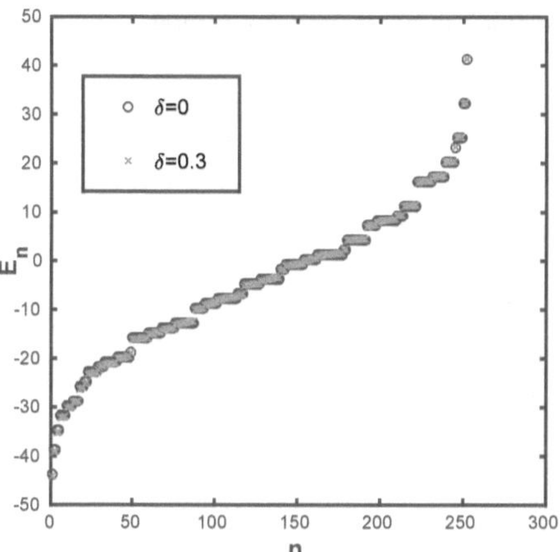

Fig. 5.6 Energy spectrum of the Haldane-Shastry model at no and weak disorder computed for 12 spins. The spectrum at weak disorder closely resembles that of the clean Haldane-Shastry model.

has volume law entanglement. The coefficient of the volume law is, however, much less than for an ergodic system. We further confirmed this non-ergodic behavior from the level spacing statistics that did not match the random matrix predictions for a thermal or a many-body localized phase.

The behavior of strongly-correlated quantum many-body systems are affected by several factors, including symmetries and the range of the terms in the Hamiltonian. It has been argued that SU(2) symmetry is not compatible with many-body localization [11, 70], and it is hence expected that the disordered Haldane-Shastry model does not many-body localize. On the other hand, low entanglement and non-ergodic behaviors, which are different from many-body localization, have been found in a nearest neighbor model with SU(2) symmetry [71, 72]. There have been several studies investigating whether many-body localization can survive in the presence of different types of long-range terms in the Hamiltonian [64, 94, 95, 65], since one might expect that long-range terms will increase the amount of entanglement. Similarly, it is unclear, whether nonlocal terms will destroy the non-ergodic, low entanglement behavior seen in local models with SU(2) symmetry. The results presented in this chapter provide an example, which shows that the non-ergodic, low entanglement behavior can also occur in nonlocal models. The obtained results motivate the search for further, interesting, non-ergodic phases in models with SU(2) symmetry and long-range interactions, as well as systematic studies to clarify under which types of conditions, models with non-ergodic behavior may appear.

Chapter 6

Topologically ordered quantum many-body scars

In Chapter 3, we saw that the weak ergodicity breaking through the mechanism of quantum many-body scars occurs due to the existence of one or more athermal eigenstates that have low entanglement embedded in a sea of ergodic states. Systematic constructions have been proposed to allow for a scar state to have nonchiral topological order [84] and survive in a disordered setting [80]. In this chapter, we provide two constructions of topologically ordered quantum scars in two parts.

In the first part, we construct a few-body Hamiltonian on 2D lattices, which has a scar state with chiral topological order in the middle of the spectrum. We also construct a few-body Hamiltonian on 1D lattices where the scar state has quantum critical properties. For both models, a parameter allows us to place the scar state at any desired position in the spectrum. The models are defined on slightly disordered lattices in 1D and 2D, and the scar state in 2D is a lattice Laughlin state. We provide evidence that the spectra are thermal by showing that the level spacing distributions are Wigner-Dyson and that the entanglement entropies of the excited states are close to the Page value. The scar state, on the contrary, is non-ergodic and has a much lower entanglement entropy. We demonstrate the topological properties of the scar state in 2D by introducing anyons into the state.

In the second part, following a general strategy, we construct a class of quantum dimer models on the kagome lattice containing quantum many-body scar states with non-chiral topological order and sub-volume entanglement. We numerically demonstrate that the remaining eigenstates in the spectrum thermalize by analyzing their level spacing statistics and entanglement entropy.

This chapter is based on the following references [96, 97].

- " *Quantum many-body scars with chiral topological order in 2D and critical properties in 1D* ", **N. S. Srivatsa**, J. Wildeboer, A. Seidel & A. E. B. Nielsen, **Phys. Rev. B 102, 235106 (2020)**

- " *Topological Quantum Many-Body Scars in Quantum Dimer Models on the kagome Lattice* ", J. Wildeboer, A. Seidel, **N. S. Srivatsa**, A. E. B. Nielsen & O. Erten, **arXiv:2009.00022**

6.1 Quantum scars with chiral topological order in 2D and critical properties in 1D

Our construction starts from sets of operators A_i and B_i that annihilate the scar state $|\Psi_{\text{scar}}\rangle$ of interest, i.e.

$$A_i|\Psi_{\text{scar}}\rangle = B_i|\Psi_{\text{scar}}\rangle = 0. \tag{6.1}$$

We use these to construct the Hamiltonian

$$H = \sum_i A_i^\dagger A_i - \gamma \sum_i B_i^\dagger B_i, \tag{6.2}$$

which has $|\Psi_{\text{scar}}\rangle$ as an exact zero energy eigenstate. The Hamiltonian H can have both positive and negative eigenvalues, and the real parameter γ can be used to adjust the position of the scar state relative to the range of the many-body spectrum. For the special case $\gamma = 0$, we observe that $|\Psi_{\text{scar}}\rangle$ is the ground state. We expect this approach to work quite generally to construct scar models, although the thermal properties of the spectrum need to be checked in each case.

6.1.1 Hamiltonians for the scar states

The quantum many-body scar state $|\Psi_{\text{scar}}\rangle$, that we wish to embed in the many-body spectrum is given by the analytical state derived from CFT described in Eq (2.4) in the Chapter 2. Specifically, we fix the number of particles in the state to be $M = N/2$ and consider the strict lattice limit with $\eta = 1$. In 1D, when the lattice sites sit evenly on a circle ($z_j = e^{i2\pi j/N}$), the state coincides with the ground state of the Haldane-Shastry model described in Eq. (2.18) after a transformation to the spin basis [45, 46]. In 2D, on a square lattice, the state is the half filled bosonic lattice Laughlin state, which has chiral topological order [4].

We now construct few-body Hamiltonians for which $|\Psi_{\text{scar}}\rangle$ is an exact eigenstate. The construction relies on the observation that the state (2.7) can be written as a correlation function in conformal field theory, and that one can use this property to derive families of operators that annihilate the state. The operators that we use below were derived in [5].

6.1.1.1 Hamiltonian in 1D

We first consider a 1D system, which is obtained by restricting all the positions z_j to be on the unit circle, i.e. $z_j = e^{i\phi_j}$ for all j, where the ϕ_j are real numbers. We also require that all the z_j are different from one another. It was shown in [5] that the operators

$$
\begin{aligned}
\Lambda_i^{\text{1D}} &= \sum_{j(\neq i)} w_{ij}[d_j - d_i(2n_j - 1)], \\
\Gamma_i^{\text{1D}} &= \sum_{j(\neq i)} w_{ij}d_i d_j, \qquad w_{ij} = \frac{z_i + z_j}{z_i - z_j},
\end{aligned}
\tag{6.3}
$$

annihilate $|\Psi_{\text{scar}}\rangle$. The operators d_j, $d_j^\dagger$, and $n_j = d_j^\dagger d_j$, acting on site j, are the annihilation, creation, and number operators for hardcore bosons, respectively.

We build the Hamiltonian for the scar model in 1D by constructing the Hermitian operator

$$
H_{\text{1D}} = \sum_i (\Lambda_i^{\text{1D}})^\dagger (\Lambda_i^{\text{1D}}) + (\alpha - 2)\sum_i (\Gamma_i^{\text{1D}})^\dagger (\Gamma_i^{\text{1D}}),
\tag{6.4}
$$

which has $|\Psi_{\text{scar}}\rangle$ as an exact energy eigenstate with energy zero. The real parameter α is chosen to position the scar state at the desired position in the spectrum of H_{1D} as we shall see below. Expanding this expression, we get

$$
H_{\text{1D}} = \sum_{i\neq j} G_{ij}^A d_i^\dagger d_j + \sum_{i\neq j} G_{ij}^B n_i n_j + \sum_{i\neq j\neq l} G_{ijl}^C d_i^\dagger d_l n_j + \sum_i G_i^D n_i + G^E,
\tag{6.5}
$$

where the coefficients are given by

$$
\begin{aligned}
G_{ij}^A &= -2w_{ij}^2, \\
G_{ij}^B &= (2 - \alpha)w_{ij}^2 + 4\sum_{l(\neq j\neq i)} w_{ij}w_{il}, \\
G_{ijl}^C &= -\alpha w_{ji}w_{jl}, \\
G_i^D &= -2\sum_{j(\neq i)} w_{ij}^2 - \sum_{j\neq l(\neq i)} w_{ij}w_{il}, \\
G^E &= \frac{-N(N-2)(N-4)}{6}.
\end{aligned}
\tag{6.6}
$$

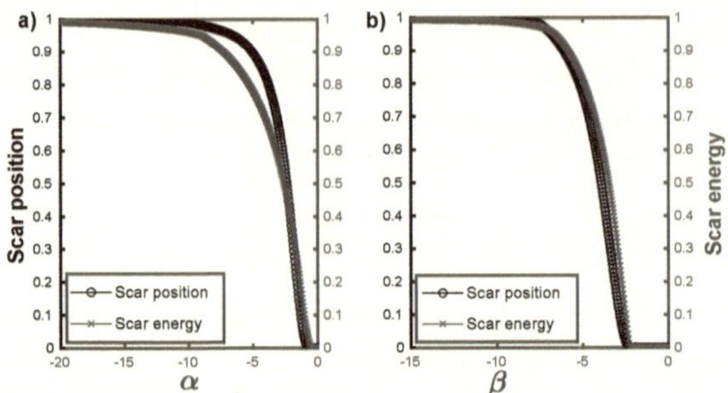

Fig. 6.1 (a) Normalized position (6.11) and normalized energy (6.12) of the scar state in the many-body spectrum of the 1D Hamiltonian (6.5) plotted as a function of the tuning parameter α for a slightly disordered lattice on a unit circle with $N = 16$ sites. (b) Same as (a), but computed for the 2D Hamiltonian (6.16) for a slightly disordered square lattice of size 4×4. It is seen that, both in 1D and 2D, the scar state can be positioned at any desired place in the many-body spectrum by a suitable choice of α and β, respectively.

Eq. (6.5) is thus a real, particle-number conserving 2-body Hamiltonian with some non-local terms. As we aim at constructing a model with a thermal spectrum, we add a small amount of random disorder to the lattice site positions to avoid any additional symmetries. Specifically, we choose

$$z_j = e^{2\pi i(j+\gamma_j)/N}, \qquad j \in \{1,2,\ldots,N\}, \tag{6.7}$$

where γ_j is a random number chosen with constant probability density in the interval $[-\frac{\delta}{2}, \frac{\delta}{2}]$ and δ is the disorder strength. In all the numerical computations below, we choose $\delta = 0.5$ and consider one particular disorder realization.

6.1.1.2   Hamiltonian in 2D

We next allow the z_j to be in the complex plane, and we assume that all the z_j are different from one another. In this case, the operators

$$\Lambda_i^{2D} = \sum_{j(\neq i)} c_{ij}[d_j - d_i(2n_j - 1)],$$

$$\Gamma_i^{2D} = \sum_{j(\neq i)} c_{ij}d_i d_j, \qquad c_{ij} = \frac{1}{z_i - z_j}, \tag{6.8}$$

annihilate $|\Psi_{\text{scar}}\rangle$ as shown in [5], and we use these to construct the Hamiltonian

$$H_{2D} = \sum_i (\Lambda_i^{2D})^\dagger (\Lambda_i^{2D}) + \beta \sum_i (\Gamma_i^{2D})^\dagger (\Gamma_i^{2D}) \tag{6.9}$$

for the scar model in 2D. In this expression, β is a real parameter, which we can adjust to position the scar state anywhere in the spectrum as we shall see below. After expanding out the terms, we get

$$H_{2D} = \sum_{i \neq j} F_{ij}^A d_i^\dagger d_j + \sum_{i \neq j} F_{ij}^B n_i n_j + \sum_{i \neq j \neq k} F_{ijk}^C d_i^\dagger d_j n_k + \sum_{i \neq j \neq k} F_{ijk}^D n_i n_j n_k + \sum_i F_i^E n_i, \tag{6.10}$$

where

$$\begin{aligned}
F_{ij}^A &= 2c_{ij}^* c_{ij} + \sum_{k(\neq i, \neq j)} (c_{ki}^* c_{kj} + c_{ji}^* c_{jk} + c_{ik}^* c_{ij}), \\
F_{ij}^B &= \beta c_{ij}^* c_{ij} - 2 \sum_{k(\neq i, \neq j)} (c_{ij}^* c_{ik} + c_{ik}^* c_{ij}), \\
F_{ijk}^C &= -2(c_{ji}^* c_{jk} + c_{ik}^* c_{ij}) + \beta c_{ki}^* c_{kj}, \\
F_{ijk}^D &= 4c_{ik}^* c_{ij}, \\
F_i^E &= \sum_{j(\neq i)} (c_{ji}^* c_{ji} + c_{ij}^* c_{ij}) + \sum_{j,k(\neq i)} c_{ij}^* c_{ik}.
\end{aligned}$$

The Hamiltonian conserves the number of particles, it is nonlocal, and it consists of terms involving up to three particles.

We disorder the positions of the lattice sites slightly to avoid additional symmetries in the model. Specifically, if z_i^c is the position of the ith site in the perfect square lattice, we choose $z_i = z_i^c + \delta(\varepsilon_a + i\varepsilon_b)$ where ε_a and ε_b are independent random numbers chosen with constant probability density in the interval $[0,1]$ and δ is the disorder strength. In all the numerical computations below, we choose $\delta = 0.1$ and consider one particular disorder realization.

6.1.2 Properties of the Hamiltonians

Since the Hamiltonians conserve the number of particles, and the scar state has $N/2$ particles, where N is the number of lattice sites, we restrict our focus to the sector with lattice filling one half in the rest of the article.

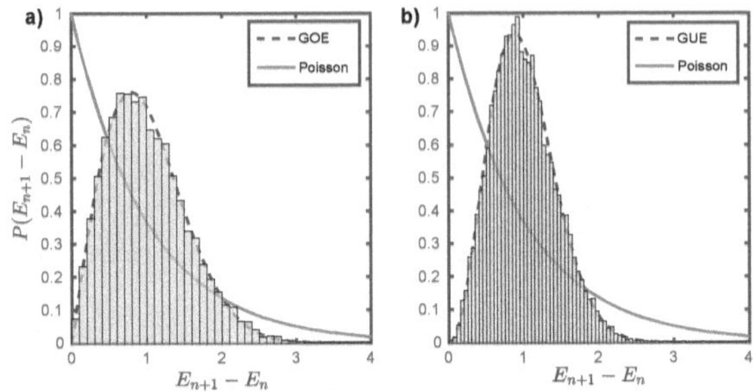

Fig. 6.2 (a) The level spacing distribution for the Hamiltonian in 1D for $N = 16$, with $\alpha = -4.5$ matches the GOE statistics for a system that is time-reversal invariant. (b) The level spacing distribution for the Hamiltonian in 2D for $N = 16$, with $\beta = -4$ matches the GUE statistics for a system that is not time-reversal invariant. This suggests that most of the states in the spectrum are ergodic.

6.1.2.1   Positioning the scar state

We first show that the parameters α and β serve as tuning parameters to place the scar state at the desired position of the many-body spectrum. If the scar state is the nth eigenstate, when ordering the eigenstates in ascending order according to energy, we define the normalized position of the scar state as

$$\text{Normalized position} = \frac{n}{D}, \tag{6.11}$$

where $D = \binom{N}{N/2}$ is the dimension of the Hilbert space in the half filled sector. The energy of the scar state is zero by construction, and the normalized energy of the scar state is hence

$$\text{Normalized energy} = \frac{-E_{\min}}{E_{\max} - E_{\min}}, \tag{6.12}$$

where $E_{\min}$ ($E_{\max}$) is the lowest (highest) energy in the spectrum of the Hamiltonian. In Fig. 6.1, we show these two quantities for the 1D and 2D Hamiltonians in Eqs. (6.5) and (6.16), respectively. From the plots it is seen that the scar state can be placed at any desired position in the many-body spectrum by choosing α and β appropriately. The scar state is the ground state for $\alpha = 0$ and $\beta = 0$, respectively.

6.1.2.2   Level spacing distribution

We next show numerically that the level spacing distributions follow the Wigner surmise for ergodic Hamiltonians [98]. In particular, time reversal invariant Hamiltonians with real

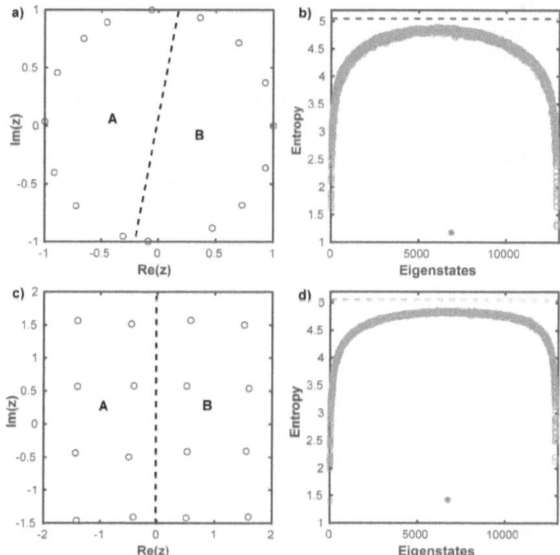

Fig. 6.3 (a,c) The slightly disordered lattices in 1D and 2D with $N = 16$ spins. The red dashed lines show the partitioning into subsystems A and B used for computing the entanglement entropies. (b,d) Entanglement entropy for all the eigenstates of the Hamiltonian in 1D [Eq. (6.5)] with $\alpha = -4.5$ and in 2D [Eq. (6.16)] with $\beta = -4$, respectively. The states in the middle of the spectrum have entropies close to the Page value $[N \ln(2) - 1]/2 = 5.05$ (dashed lines), except the scar state (marked with a red star), which has a much lower entropy.

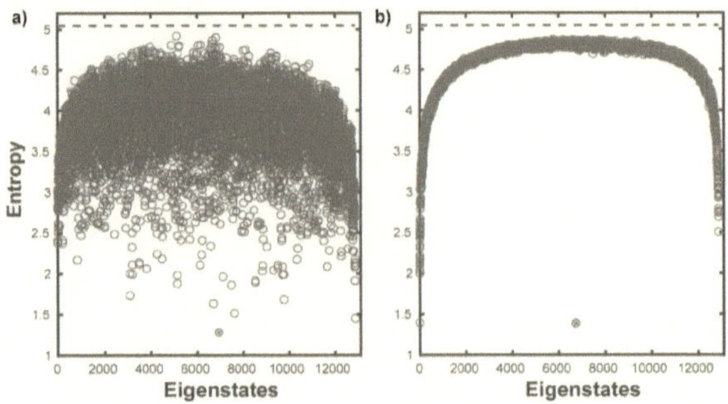

Fig. 6.4 (a) Entanglement entropy for all the eigenstates of the Hamiltonian in 1D [Eq. (6.5)] with $\alpha = -4.5$ and $\delta = 0$. (b) Entanglement entropy for all the eigenstates of the Hamiltonian in 2D [Eq. (6.16)] with $\beta = -4$ and $\delta = 0$. Dashed lines indicate the Page value $[N\ln(2) - 1]/2 = 5.05$ and the scar state is marked with a red star.

matrix entries are known to follow Gaussian orthogonal ensemble (GOE) statistics while those with complex entries breaking time reversal symmetry are known to follow Gaussian unitary ensemble (GUE) statistics [99]. In Fig. 6.2, we show the level spacing distributions for the Hamiltonians in 1D and 2D using the standard technique of unfolding the energy spectrum [100]. The energy level spacing distribution follows GOE statistics for the 1D case while it follows GUE statistics for the 2D case. This suggests that most of the states in the many-body spectrum of the Hamiltonians are ergodic.

6.1.2.3   Entanglement entropy

Scar states are non-ergodic states that reside in the bulk of the spectrum where the density of states is high. The entanglement entropy of scar states is low compared to the entanglement entropy of the states in the middle of the spectrum that obey the eigenstate thermalization hypothesis. When the subsystem is half of the system, the entanglement entropy of the latter states is close to the Page value $[N\ln(2) - 1]/2$ [57].

We now show that the models considered here display such a behavior. Specifically, we consider the von Neumann entropy $S = -\text{Tr}[\rho_A \ln(\rho_A)]$ of a subsystem A, where $\rho_A = \text{Tr}_B(|\psi\rangle\langle\psi|)$ is the reduced density matrix after tracing out part B of the system and $|\psi\rangle$ is the energy eigenstate of interest. For the 1D case, we choose region A to be half of the chain, and for the 2D case, we choose region A to be the left half of the lattice. The choice of subsystems and the entanglement entropy for all the states for a disordered case are shown in Fig. 6.3. It is seen that the states in the middle of the spectrum have large entanglement

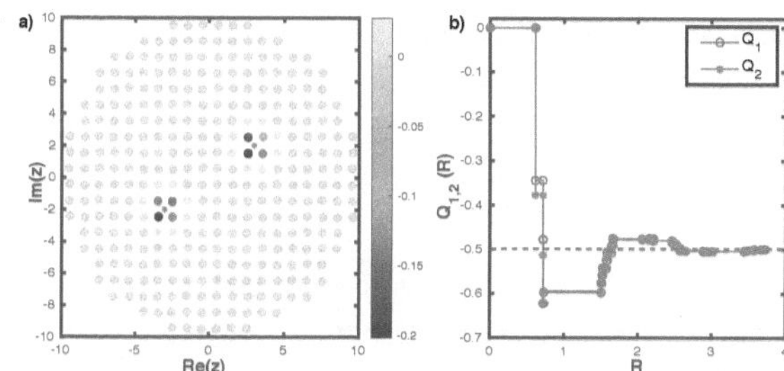

Fig. 6.5 (a) Density difference profile $\rho(z_i)$ [see (??)] on a slightly disordered square lattice with the two quasihole coordinates, w_1 and w_2, marked by the blue and the red star, respectively. The observation that $\rho(z_i)$ is only nonzero in a finite region around each of the quasihole coordinates shows that the anyons are well screened. (b) The excess densities, Q_1 and Q_2 [see (6.13)], as a function of the radius R converge to the correct value -0.5 for a quasihole, which shows that the state is topological (see text). The number of lattice sites is $N = 316$ and the edge of the lattice is chosen to be roughly circular in order to resemble a quantum Hall droplet. The error bars due to Metropolis sampling are of order 10^{-4}.

entropy close to the Page value while the scar state has low entanglement entropy in both 1D and 2D.

We also compute entropies for models without disorder as shown in Fig. 6.4. For the 1D case, the entropies in the middle of the spectrum are no longer distributed around the Page value. However, for the 2D clean model, the entropies close to the middle of the spectrum are close to the Page value while the scar state has low entanglement.

6.1.3 Properties of the scar states

As mentioned above, the scar state coincides with the ground state of the HS model, when defined on a uniform 1D lattice, and on a 2D square lattice, the scar state is a lattice Laughlin state. In Chapter 4, we showed numerically that the HS state stabilizes Luttinger liquid behavior even after introducing disorder. Hence, for the 1D case, and on a slightly disordered lattice, the scar state is quantum critical which shows logarithmic divergence of entanglement entropy with subsystem size as described in Eq. (4.9).

The lattice Laughlin state (2.4) is known to have chiral topological order and to support anyonic excitations when defined on different regular lattices [4, 5, 101, 34]. Here, we confirm the topological properties of the scar state on a slightly disordered lattice by introducing anyons. The anyons are introduced by modifying the state through Eq. (2.13) as discussed in Chapter 2. For a topological liquid, one expects a constant density of particles in the bulk. In the presence of quasiholes, however, the density is reduced in a local region around each of the quasihole coordinates w_k. We define the excess density of the kth quasihole as

$$Q_k = \sum_{i=1}^{N} \rho(z_i)\,\Theta(R - |z_i - w_k|), \quad k \in \{1,2\}, \tag{6.13}$$

where $\rho(z_i)$ is given by Eq. (2.15). For the case considered here, we specifically have $q = 2$, $S = 2$ and $p_j = 1\ \forall j$. The Heaviside step function $\Theta(\ldots)$ restricts the sum to a circular region of radius R and center w_k. When R is large enough to enclose the quasihole and small enough to not enclose other quasiholes or part of the edge of the lattice, the excess density is equal to minus the charge of the quasihole. The charge of the quasihole is $1/2$ for the considered Laughlin state. We use Metropolis-Hastings algorithm to compute $\rho(z_i)$ and Q_k in Fig. 6.5. The density difference profile shows that the quasiholes are well screened, and we find that the excess densities Q_1 and Q_2 converge to -0.5 as we increase the radius R. For the lattice Laughlin state, it can be shown analytically that the quasiholes have the correct braiding properties, if it can be assumed that the anyons are screened and have the correct charge [102]. Since the scar state hence admits the insertion of local defects with anyonic properties, it follows that it is a topologically ordered state even in the presence of the amount of disorder considered here.

6.2 Scars with non-chiral topological order in quantum dimer models

In this section, we will discuss yet an other construction of a model with quantum many-body scars with non-chiral topological order. We will begin with the description of the quantum dimer model on a kagome lattice and then show how we turn them into models containing scars. Finally, we will discuss the computations of level spacing statistics and entanglement entropies and argue that the model is generic and the non-scar states are thermal.

6.2.1 Quantum dimer model on a kagome lattice

Quantum dimer models were originally modelled to study the physics of high-temperature superconductors through Anderson's resonating valence bond physics [103]. The quantum dimer model is defined on a Hilbert space of distinct orthonormal states representing the allowed hard-core dimer coverings on the lattice. In this formulation, a dimer represents a bond which is an SU(2) singlet between spins at its endpoints.

In this section, we will consider the dimer model on the kagome lattice [104] which can be graphically represented by the following Hamiltonian.

$$\mathcal{H} = \sum_{\hexagon} \Bigg[-t_1 \left(\big| \cdots \big\rangle \big\langle \cdots \big| + \big| \cdots \big\rangle \big\langle \cdots \big| \right) + V_1 \left(\big| \cdots \big\rangle \big\langle \cdots \big| + \big| \cdots \big\rangle \big\langle \cdots \big| \right)$$
$$+ \sum_{r_{rot}} \Big[-t_2 \left(\big| \cdots \big\rangle \big\langle \cdots \big| + \big| \cdots \big\rangle \big\langle \cdots \big| \right) + V_2 \left(\big| \cdots \big\rangle \big\langle \cdots \big| + \big| \cdots \big\rangle \big\langle \cdots \big| \right) \Big]$$

$$+ \cdots$$

$$-t_8 \left(\big| \cdots \big\rangle \big\langle \cdots \big| + \big| \cdots \big\rangle \big\langle \cdots \big| \right) + V_8 \left(\big| \cdots \big\rangle \big\langle \cdots \big| + \big| \cdots \big\rangle \big\langle \cdots \big| \right) \Bigg] \quad (6.14)$$

The above Hamiltonian consists of "kinetic" terms, t that allow for a local rearrangement of a small number of dimers and "potential" terms, V, that are diagonal in the dimer basis which correspond to local arrangements of dimers. The dimers are represented by the Magenta bonds and sum is over all 12-site star plaquettes of the lattice. All the kinetic terms execute resonance moves along one of 32 different loops contained within the star. The occupied links alternate along the loop and each move changes the occupancy along the loop. The list of different possible loops is described in the Table 6.1.

The model is completely integrable at the special Rokhsar-Kivelson (RK) point when $t_1 = \cdots = t_8 = V_1 = \cdots = V_8 > 0$ and the ground state is the equal amplitude superposition of all the permissible dimer coverings given by,

$$|\Psi\rangle = \sum_D |D\rangle . \quad (6.15)$$

For the case of periodic boundary conditions, on a torus, the sum needs to be restricted to one topological sector which leads to a 4-fold degenerate state. Topological sectors may be classified through the winding numbers W_x and W_y. To determine $W_x(W_y)$, one considers a horizontal (vertical) line around the torus which intersects the links and $W_x(W_y)$ is the parity of the number of dimers intersected. On the Kagome lattice, this model is fully integrable at the RK-point with the sums of operators in Eq (6.14) for any given star commutes for different stars [104]. Further, this RK-point lies in the interior of $\mathbb{Z}_2$ topological phase [104–106] and the topological order has been tested numerically through the topological entanglement entropy computed for the state in Eq. (6.15) [107].

12-site star								
loop type A_i	A_1	A_2	A_3	A_4	A_5	A_6	A_7	A_8
# of links	6	8	8	10	8	10	10	12
r_{rot}	1	3	6	3	6	6	6	1

Table 6.1 A list of all eight possible loops surrounding the central hexagon within a star-shaped cell, up to rotational symmetry. The total number of rotationally related loops r_{rot} is shown for completeness (e.g., $r_{\mathrm{rot}} = 1$ for the hexagon and the star and $r_{\mathrm{rot}} > 1$ for the other shapes). Each dimerization realizes exactly one of the resulting 32 loops, where the links of the loop alternate between occupied and unoccupied, yielding two possible realizations via dimers for each loop.

6.2.2 Kagome dimer model with scars

We now construct yet an other quantum dimer model on the kagome lattice which admits quantum scar states in the spectrum. The construction begins with the fact that the state in Eq .(6.15) are also annihilated by the Hamiltonian in Eq .(6.14) whenever $t_i = V_i$. This is because each local term of the Hamiltonian associated with $t_i = V_i$ annihilates the state described in Eq .(6.15). Importantly, with this new choice, the frustration free Hamiltonian is no longer integrable but the state described in Eq .(6.15) is still its exact zero energy eigenstate. With this, we introduce $t_i = V_i = \alpha_i$ and choose the α_i different and not all of the same sign. Crucially, now the state described in Eq .(6.15) is still a zero energy eigenstate of the Hamiltonian, but not its ground state anymore. Rather, it is an eigenstate that lives somewhere in the middle of the spectrum and hence a quantum many-body scar.

The Hamiltonian for the dimer model containing the scar can be expressed as follows

$$\mathcal{H}^{scar} = \sum_{\bigstar} \sum_{l=1}^{32} \alpha_l \left(|D_l\rangle - |\overline{D_l}\rangle \right) \left(\langle D_l| - \langle \overline{D_l}| \right) . \tag{6.16}$$

The sums in (6.16) run over all 12-site kagome stars and over all 32 loop coverings. D_l and $\overline{D_l}$ represent the dimerizations associated with loop l. This strategy could be followed easily while preserving all lattice symmetries. It is also important to restrict oneself within the symmetry sectors to compute level spacing statistics as there is no level repulsion between different sectors. Thus, to avoid an over-abundance of symmetry sectors, we preserve only

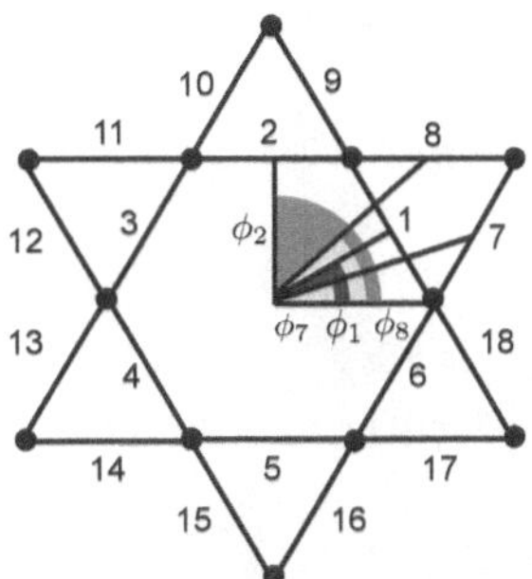

Fig. 6.6 Labelling convention for the links of the 12-site star-shaped cells of the kagome lattice. The 12-site kagome star consists of a total of 18 links. Internal links that make up the central hexagon are labeled by link indices from 1 to 6, while links $7, \ldots, 18$ refer to external links. Internal (external) links are associated with internal (external) angles. We show the internal angles $\phi_1 = \pi/6$ and $\phi_2 = \pi/2$ associated with the links 1 and 2, respectively. External angles are also illustrated and differ from adjacent internal ones by $\pm\frac{\pi}{12}$, e.g. $\phi_7 = \phi_1 - \frac{\pi}{12}$ and $\phi_8 = \phi_1 + \frac{\pi}{12}$.

translational symmetry by choosing

$$\alpha_l = C + \sum_{l' \in \text{loop}} \sin\left(5 \cdot \phi_{l'} + \delta\right) . \tag{6.17}$$

which simulates the influence of a substrate with 5-fold rotational symmetry. The ϕ_l's are angles associated with each link that the respective loop covers. They are defined to be

$$\phi_j = \begin{cases} (2j-1)\frac{\pi}{6} & \text{for } j = 1,\ldots,6 \\ \phi_{\frac{j-5}{2}} - \frac{\pi}{12} & \text{for } j = 7,9\ldots,17 \\ \phi_{\frac{j-6}{2}} + \frac{\pi}{12} & \text{for } j = 8,10\ldots,18, \end{cases} \tag{6.18}$$

as shown in Figure 6.6. We choose $C = -0.05$ to make dimer-loops with inversion symmetry contribute, and $\delta = 0.1$ to render the mirror axes of the "substrate" different from those of the lattice.

While the scar state of the model and its properties are analytically under control, we proceed by numerically investigating the genericity of its other levels in the following sections.

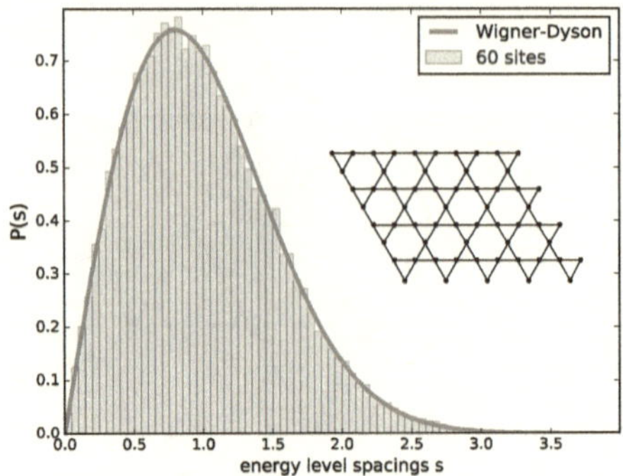

Fig. 6.7 Distribution of energy levels for time reversal invariant zero-momentum, $(W_x, W_y) = (0,0)$ topological sector, which contains a scar state (6.15). The inset shows the 60-site kagome lattice used in the calculation. An unfolding technique using 4378 groups containing 12 energies each has been used for binning the data (cf., e.g., Ref. [?]). The resulting data closely resemble a GOE distribution (solid curve), indicating that almost all states thermalize.

6.2.2.1 Level spacing statistics

To test whether the non-scar states conform to the ETH, we compute the level spacing statistics discussed in Chapter. 4. We focus on (translational) symmetry sectors with time reversal symmetry, which contain the scar state. Fig 6.7 shows the distribution of energy eigenvalues for a 60 site kagome lattice with PBCs, within the zero-momentum sector that has the scar state located roughly in the middle of the spectrum (see Fig 6.8). Here, we work within the $(W_x, W_y) = (0,0)$ topological sector and use an unfolding technique to bin the data [100]. The distribution is well described by the Gaussian orthogonal ensemble (GOE), as expected for generic real matrices. Further, we compute the adjacent gap ratios, $r_n = s_n/s_{n-1}$ and $\tilde{r}_n = \min(r_n, 1/r_n)$, where $s_n = E_n - E_{n-1}$. We find the average of the quantities $\tilde{r}_n$ over the energy slice described, and over all symmetry-inequivalent time-reversal invariant momentum sectors within the $(W_x, W_y) = (0,0)$ topological sector to be $\langle \tilde{r} \rangle = 0.5333$. This is quite close to the exact value of $\tilde{r}_{\text{GOE}} = 0.5359$ [56], and markedly different from the corresponding value $\tilde{r}_{\text{Poisson}} = 0.3863$ for the Poisson distribution [89]. This lends strong support to the ETH that the majority of the high energy states in the spectrum of Hamiltonian (6.16) thermalize.

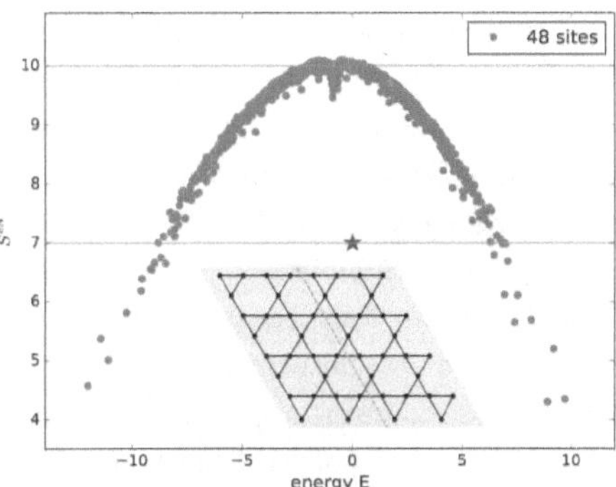

Fig. 6.8 The von Neumann entanglement entropy for all states within the zero momentum sector of topological winding numbers $(0,0)$ for a 48 site kagome lattice, bi-partitioned into two 24-site "ribbons" (inset). The scar state has $S^{vN} = 7$ and is marked by a blue star. Thermalizing eigenstates of similar energy are well separated and have $S^{vN} \approx 10$.

6.2.2.2 Entanglement entropy

To complement the above findings, we calculate bipartite entanglement entropy for all states of the scar-containing symmetry sector (fixing also the topological sector) for a 48 site kagome lattice with PBCs. By their definition, quantum many-body scar states belong to the bulk of the spectrum while simultaneously violating the ETH, i.e., they fail to thermalize and display low (sub-volume) entanglement behavior. In contrast, generic high-energy states do thermalize and exhibit a volume-dependent entanglement behavior. For the ribbon described, whose boundary passes eight unit cells on each side, and in the presence of the topological sector constraint, one may show that the (base 2) von Neumann entanglement entropy, $S^{vN} = -\sum_w w \log_2 w$, equals 7. Here, the sum goes over the eigenvalues of the local density matrix of the ribbon. Fig. 6.8 clearly shows that the scar state (blue star) is isolated from the rest of the spectrum (purple dots) in terms of its much lower entanglement as compared to surrounding bulk energy eigenstates. This confirms the state (6.15) as a legitimate quantum many-body scar.

6.3 Conclusion

In this chapter, we provided constructions of two rather different models for quantum many-body scars with topological order.

In the first section, we built 2D models featuring quantum many-body scar states that have chiral topological order. Specifically, we have used sets of operators that annihilate the scar state of interest to construct nonlocal, few-body Hamiltonians, which have the scar state as an exact eigenstate. Using a similar construction for a 1D system, we have also obtained a scar model in which the scar state is critical. For both models, a parameter allows us to place the scar state at any desired position in the spectrum. We have found that the level spacing distributions of the spectra are Gaussian and that the excited states in the middle of the spectra have entanglement entropies close to the Page value, which is what is expected for thermal spectra. The scar state, however, is special and has a much lower entanglement entropy. Finally, we have confirmed the topological nature of the scar state in 2D by showing that one can insert anyons with the correct charge and braiding properties into the state.

In the second section, we demonstrated a general approach to turn classes of frustration free lattice Hamiltonians into ones containing isolated quantum many-body scars in their spectrum while retaining most or all symmetries. We applied this strategy to a two-dimensional quantum dimer model on the kagome lattice, retaining full translational symmetry. We demonstrated that this model contains an exactly known quantum many-body scar with analytically accessible entanglement properties. We established that the remainder of the eigenstates and energy spectrum exhibit no "fine-tuned" behavior. Specifically, for a 60-site kagome lattice, we showed that bulk energies conform to the Gaussian ensembles expected for their respective symmetry sectors, and we calculated von Neumann entanglement entropies for all states within the scar-sector of a 48 site kagome lattice, exposing the scar's isolated character.

Chapter 7

Taming the non-local lattice Hamiltonians constructed from CFT

As we briefly discussed in Chapter 2, CFT has turned out to be a useful tool to construct lattice versions of fractional quantum Hall physics in 2D and critical properties in 1D. Hence, it provides an alternative route to obtain FQH physics in lattice systems, compared to fractional Chern insulators. One advantage is that the Hamiltonians by construction have states with a particular topology as ground states. Interestingly, in the previous chapter, we saw that the CFT construction also allows for realizing models with scars that are lattice FQH states.

We recall that the parent Hamiltonians are given by the expressions of the form

$$H_{\text{Exact}} = \sum_i \Lambda_i^\dagger \Lambda_i, \tag{7.1}$$

where $\Lambda_i|\Psi\rangle = 0$ and $|\Psi\rangle$ is the lattice version of either the fractional quantum Hall state in 2D or the many-body state with critical properties. The parent Hamiltonians H_{Exact}, although few body, are non-local and for the following reasons, calls for methods to make them tractable. Firstly, the long ranged interactions are difficult to engineer in experiments. Secondly, it would be natural for the terms of a local Hamiltonian to depend only on the local lattice structure which leads to a well defined thermodynamic limit. Thirdly, the local Hamiltonians are easier to work with numerically, e.g. when studying the systems with exact diagonalization or tensor networks [108].

In the first part of this chapter, we discuss a general way to arrive at local FQH lattice Hamiltonians constructed from CFT. This method as we will emphasise, does not rely on optimization, choice of lattice or symmetries in the model. We discuss how we test this approach on two models with bosonic Laughlin-like ground states with filling factor 1/2 and 1/4, respectively. For the half filled model on the torus, we show that there is a near twofold

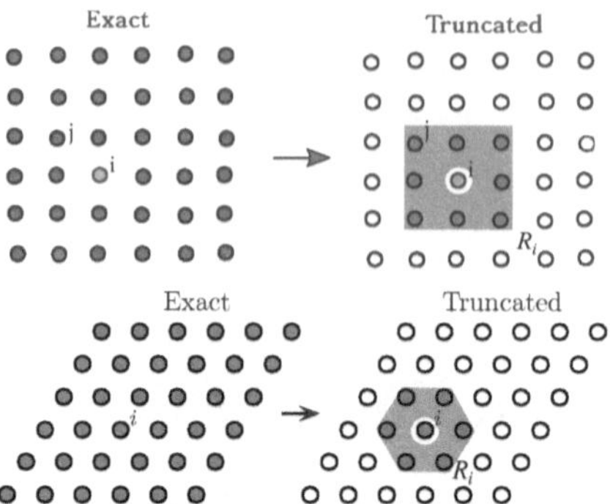

Fig. 7.1 Illustration of the truncation procedure of the Λ_i operator for a square lattice (top panel) and a triangular lattice (bottom panel). The left part of the diagram shows the exact Λ_i operator, which contains terms involving the ith lattice site (shown in green) and any of the other lattice sites (shown in red). The right part of the diagram displays the truncated form of the Λ_i operator, which only contains terms involving the ith lattice site (shown in green) and any of the sites within a local region (shown in red with a blue background).

degeneracy with a gap to the first excited state as one would expect as a consequence of the topological order. We demonstrate that the overlap per site is higher than 0.99 for the two lowest energy states for all cases considered. For the quarter filled model, we show that a larger truncation radius is needed to get high overlaps for low energy states on small lattices.

In the second part, we discuss how we arrive at local models in 1D whose ground states have similar critical properties as the exact versions. Based on computations of wavefunction overlaps, entanglement entropies, and two-site correlation functions for systems of upto 32 sites, we show that the ground states of local models are close to the ground states of the exact models. There is also a high overlap per site between the lowest excited states for the local and the exact models, although the energies of the low-lying excited states are modified to some extent for the system sizes considered. We also briefly discuss possibilities for realizing the local models in ultracold atoms in optical lattices.

This chapter is based on the following references [109, 110].

- " *Truncation of lattice fractional quantum Hall Hamiltonians derived from conformal field theory* ", D. K. Nandy*, **N. S. Srivatsa*** & A. E. B. Nielsen, **Phys. Rev. B 100, 035123 (2019)** [* authors contributed equally to this work]

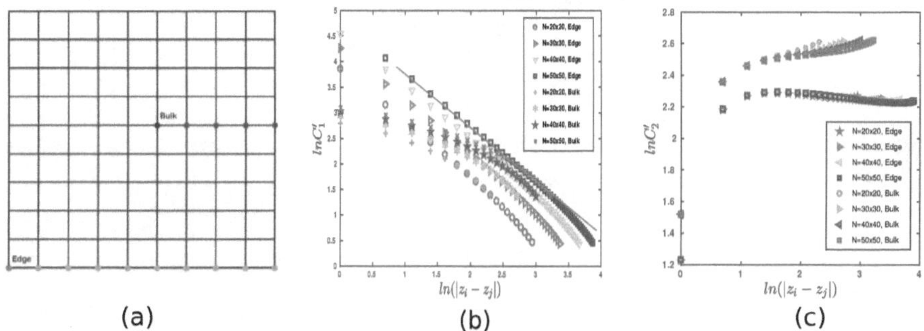

(a)   (b)   (c)

Fig. 7.2 (a) To investigate the behavior of the two-body terms in the Hamiltonian, we consider the coupling strengths between the marked sites. The term 'edge' refers to that the coupling strength is computed between the red site (site i) and each of the green sites (site j), and the term 'bulk' refers to that the coupling strength is computed between the black site (site i) and each of the blue sites (site j). (b) Log-log plot showing the behavior of $C_1' = \sum_{k(\neq i, \neq j)} \left(w_{ki}^* w_{kj} + w_{ji}^* w_{jk} + w_{ik}^* w_{ij} \right)$ related to the coefficient of the $d_i^\dagger d_j$ term with respect to distance $|z_i - z_j|$. The straight line (shown in red) has slope -1. (c) Log-log plot showing the behavior of $C_2' = 2 \sum_{k(\neq i, \neq j)} \left(w_{ij}^* w_{ik} + w_{ik}^* w_{ij} + w_{ji}^* w_{jk} + w_{jk}^* w_{ji} \right)$ related to the coefficient of the $n_i n_j$ term with respect to distance $|z_i - z_j|$.

- " *Local Hamiltonians for one-dimensional critical models* ", D. K. Nandy, **N. S. Srivatsa** & A. E. B. Nielsen, **J. Stat. Mech. 2018, 063107 (2018)**

7.1   Truncation of lattice FQH Hamiltonians

The FQH states have correlations that decay exponentially with distance in the bulk. This suggests that it should, in fact, not be necessary to have long-range interactions in the Hamiltonians. In previous work [7, 111] local Hamiltonians for a bosonic Laughlin state and a bosonic Moore-Read state with SU(2) symmetry were achieved using the following procedure. First, all terms except a selection of local terms are removed from the Hamiltonian. The coefficients of the remaining terms are then adjusted so that all terms that only differ by a lattice translation have the same strengths. Finally, the relative strengths of terms that do not only differ by a lattice translation are determined by a numerical optimization of the overlap between the CFT states and the ground states of the truncated models. The latter step involves exact diagonalization and is done for small system sizes. This way of truncating the Hamiltonian is easier for models that have SU(2) symmetry, since the symmetry reduces the number of possible terms in the Hamiltonian and hence also the number of coefficients that

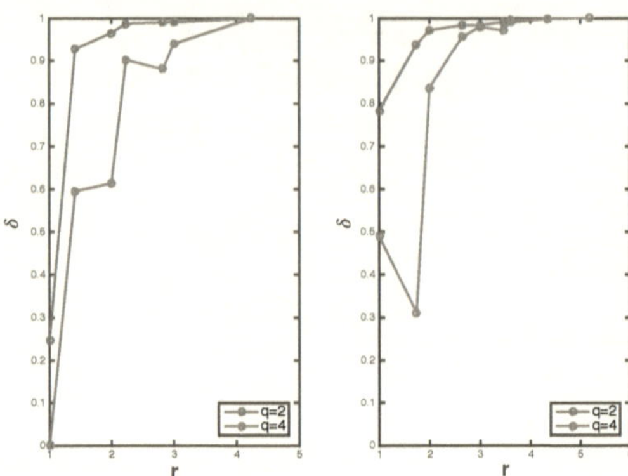

Fig. 7.3 Overlap δ versus truncation radius r computed for the case of a 4×4 square (left) or triangular (right) lattice with an open boundary.

need to be optimized. In models that do not possess such symmetries, however, many more terms in the Hamiltonian are possible, and this makes it difficult to find a suitable choice of parameters, which gives a high overlap per site independent of the system size [111].

In this section, we will discuss a different procedure to truncate Hamiltonians derived from CFT. The procedure is more general and can be applied to Hamiltonians with or without symmetries on all lattice geometries. The idea is to truncate the operator Λ_i in the first place and then proceed to construct the Hamiltonian as in Eq. (7.1). There are more advantages of this approach. The Λ_i operator is simpler to truncate than the Hamiltonian, since it contains fewer terms, and since the coefficients of the terms depend only on the relative positions of the involved sites and not on the rest of the lattice. This means that the local terms of the resulting Hamiltonians are independent of the size of the lattice used to compute them. Another advantage is that it is clear how to obtain Hamiltonians with either periodic or open boundary conditions as desired. Finally, there is no optimization involved, and once the lattice has been chosen, the only input to the procedure is the truncation radius for the Λ_i operator.

7.1.1   Exact model and wavefunctions

In the CFT construction of the Hamiltonian, null fields are used to derive a family of operators that annihilate the considered state. These operators are of increasing complexity, and for the lattice Laughlin states the three simplest are [5]

$$\Upsilon = \sum_{j=1}^{N} d_j, \tag{7.2}$$

$$\Omega = \sum_{j=1}^{N} (qn_j - 1), \tag{7.3}$$

$$\Lambda_i = \sum_{j \neq i} w_{ij}[d_j - d_i(qn_j - 1)]. \tag{7.4}$$

Here, d_j is the hardcore bosonic (fermionic) particle annihilation operator at site j for q even (odd), and $w_{ij} = \frac{1}{z_i - z_j}$. If we construct the Hamiltonian from Υ or Ω, we do not obtain a unique ground state. We therefore construct the Hamiltonian from Λ_i using Eq. (??). After multiplying out all the factors, we have

$$H^{\text{Exact}} = \sum_{i \neq j} C_1(i,j) d_i^\dagger d_j + \sum_{i \neq j} C_2(i,j) n_i n_j + \sum_{i \neq j \neq k} C_3(i,j,k) d_i^\dagger d_j n_k + \sum_{i \neq j \neq k} C_4(i,j,k) n_i n_j n_k$$
$$+ \sum_{i} C_5(i) n_i,$$

$$\tag{7.5}$$

where the coefficients are given by

$$C_1(i,j) = 2w_{ij}^* w_{ij} + \sum_{k(\neq i, \neq j)} (w_{ki}^* w_{kj} + w_{ji}^* w_{jk} + w_{ik}^* w_{ij}),$$

$$C_2(i,j) = (q^2 - 2q) w_{ij}^* w_{ij} - q \sum_{k(\neq i, \neq j)} (w_{ij}^* w_{ik} + w_{ik}^* w_{ij}),$$

$$C_3(i,j,k) = -q(w_{ji}^* w_{jk} + w_{ik}^* w_{ij}),$$

$$C_4(i,j,k) = q^2 w_{ik}^* w_{ij},$$

$$C_5(i) = \sum_{j(\neq i)} (w_{ji}^* w_{ji} + w_{ij}^* w_{ij}) + \sum_{j,k(\neq i)} w_{ij}^* w_{ik}.$$

This Hamiltonian is not SU(2) invariant. It conserves the number of particles, and we shall assume throughout that the number of particles is fixed to N/q.

We denote the unique ground state of this Hamiltonian by $|\Psi_{\text{Exact}}^{\text{P}}\rangle$, which is given by Eq. (2.4) with $\eta = 1$ (since we are in the strict lattice limit) and for any choice of lattice in the plane (i.e. for any choice of z_j). The superscript P in the wavefunction denotes that the wavefunction is defined on a plane. We will be focussing our attention to half filled ($q = 2$) and quarter filled ($q = 4$) cases.

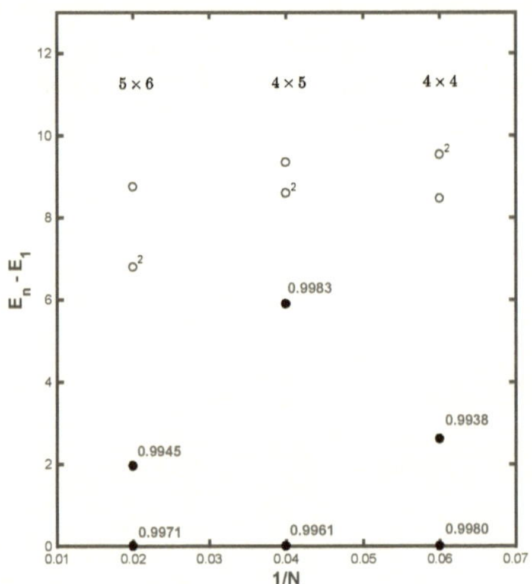

Fig. 7.4 Energy spectrum for different lattice sizes ($N = L_x \times L_y$) on the torus for $q = 2$ on a square lattice with $r = \sqrt{2}$. The energy values painted black have a high overlap per site with the exact analytical states (the overlap per site is written in purple close to the eigenstate). The black integers give the degeneracies when different from 1.

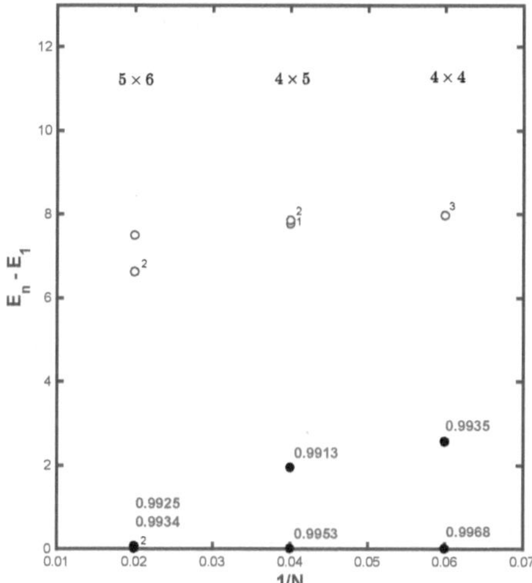

Fig. 7.5 Energy spectrum for different lattice sizes ($N = L_x \times L_y$) on the torus for $q = 2$ on a triangular lattice with $r = 1$. The energy values painted black have a high overlap per site with the exact analytical states (the overlap per site is written in purple close to the eigenstate). The black integers give the degeneracies when different from 1.

7.1.2 Behavior of the coefficients of the Hamiltonian

There are five different types of operators in the Hamiltonian: a one-body term, two two-body terms, and two three-body terms with different coupling coefficients. The one-body term is local, since each term acts only on a single site. The coefficients C_3 and C_4 of the three-body terms do not involve any summation over indices. They hence depend only on the relative positions of the involved lattice sites, and their behavior is the same everywhere on the lattice. The C_3 coefficient has two terms both of which have power law decay, while the C_4 coefficient has one term which has a power law decay with distance. It is also seen that the decaying behavior of the three-body coefficients is independent of the system size.

The coefficients of the two-body terms do, however, have a more complicated behavior. C_1 and C_2 both contain two terms. The first decays as the square of the distance between the sites, and the second is a sum. Due to these sums, the coefficients depend on the positions of all the lattice sites. We demonstrate the behavior explicitly for the case of a square lattice in Fig. 7.2. In this figure, we plot only the term containing the sum, and for C_2 we plot the sum of the contributions from $n_i n_j$ and $n_j n_i$, since these two operators are the same. We consider two cases. In the first case, we fix i to be an edge lattice point and then let j run along the edge. In the second, we fix i to be a lattice point in the middle of the lattice and then let j run towards the edge. The two cases are illustrated in Fig. 7.2(a).

From the results in Fig. 7.2(b) and 7.2(c), we make the following observations. First, the coefficients can be large even at distances comparable to the size of the lattice. Second, the behavior of the coefficients with distance depends on where the sites are on the lattice. Third, there is, in general, not a well-defined thermodynamic limit. In particular, this means that if we truncate the Hamiltonian directly, the strengths of the local terms in the bulk will not only depend on the local lattice geometry, but on the size and shape of the whole lattice. This problem does not arise when truncating the Λ_i operators instead.

7.1.3 Local model

The main purpose here is to discuss a general way to obtain a local Hamiltonian starting from the exact model. As discussed earlier, we start by truncating the operator Λ_i to obtain the local operator

$$\Lambda_i^{(\mathrm{L})} = \sum_{j \in R_i} w_{ij} \left[d_j - d_i(qn_j - 1) \right] . \tag{7.6}$$

Here, the sum is taken over the sites within some local region around the ith site as illustrated in Fig. 7.1. A natural choice is to let the local region contain all sites that are at most a distance r away from the ith site. In that case, $R_i = \{ j \mid 0 < |z_i - z_j| \leq r \}$. The truncated Hamiltonian is then constructed as

$$H^{\mathrm{Local}} = \sum_{i=1}^{N} \Lambda_i^{(\mathrm{L})\dagger} \Lambda_i^{(\mathrm{L})} . \tag{7.7}$$

The exact model is defined for open boundary conditions. This means that all the $\Lambda_i^{(\mathrm{L})}$ operators in the bulk have the same form, but the $\Lambda_i^{(\mathrm{L})}$ operators on the edge are different, because there are fewer neighbors. It is, however, straightforward to modify the truncated model to have periodic boundary conditions. We just need to use the bulk form of $\Lambda_i^{(\mathrm{L})}$ for all lattice sites and let the lattice wrap around the torus. Periodic boundary conditions reduce the finite size effects on the relatively small lattices we can investigate with exact diagonalization. In the following, we consider the truncated model for both open and periodic boundary conditions on square and triangular lattices.

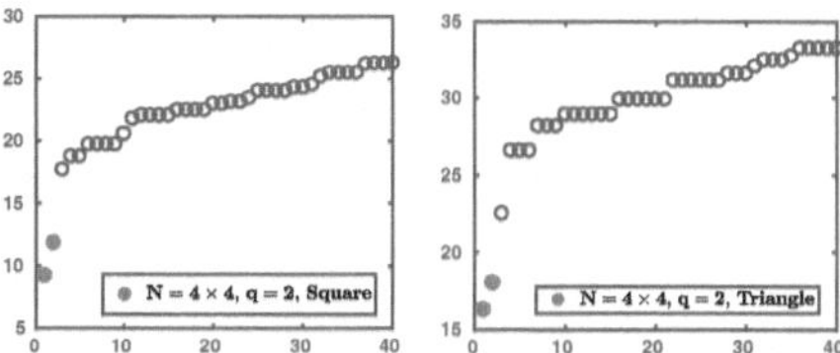

Fig. 7.6 The first 40 energy eigenvalues computed for $q = 2$ on a 4×4 square lattice with $r = \sqrt{2}$ (left) and a 4×4 triangular lattice with $r = 1$ (right). Only the states that have the same momentum quantum numbers as the analytical states can have a nonzero overlap, and the red points mark the q lowest of these states.

7.1.4 Overlap of the eigenstates

In order to quantify how well the constructed local models reproduce the exact states, we compare the eigenstates of H^{Local} computed from exact diagonalization to the exact states. On the plane, we define the overlap between the ground state $|\Psi_{\text{Local}}\rangle$ and the exact state (2.4) to be

$$\delta = |\langle \Psi_{\text{Local}} | \Psi_{\text{Exact}}^{\text{P}} \rangle|^2. \tag{7.8}$$

In Fig. 7.3, we plot the overlap δ as a function of truncation radius r for the half ($q = 2$) and quarter ($q = 4$) filled cases on a 4×4 square and triangular lattice. For the half filled case, a truncation radius of $\sqrt{2}$ on the square lattice and $\sqrt{3}$ on the triangular lattice is enough to get an overlap higher than 0.9. For the quarter filled case, a truncation radius of $\sqrt{5}$ on the square lattice and $\sqrt{7}$ on the triangular lattice is needed to get an overlap higher than 0.9. Note that the number of particles per site is a factor of two smaller for $q = 4$ than for $q = 2$, so the same number of particles will take part in the local interactions if the truncation radius is about a factor of $\sqrt{2}$ larger for $q = 4$. The behavior of the curves in Fig. 7.3 further suggests that the needed truncation radius to some extent depends on the shape of the local region.

We now consider the case of periodic boundary conditions. Because of the topological order, there are q states in total on the torus, which are given in the Appendix C. We therefore define the overlap between the jth eigenstate $|E_j\rangle$ of H^{Local} and the exact state to be

$$\Delta = \sum_{l=0}^{q-1} |\langle E_j | \tilde{\Psi}_{\text{Exact}}^{\text{T},l} \rangle|^2, \tag{7.9}$$

where $|\tilde{\Psi}_{\text{Exact}}^{\text{T},l}\rangle$ are the states in (C.1) after Gram-Schmidt orthonormalization. Due to the exponential growth of the Hilbert space dimension with system size, we expect the overlap to show an exponential decay with system size. This motivates us to also consider the overlap

Fig. 7.7 The division of the lattices used for computing the topological entanglement entropy. The different regions are marked with colors. The two upper plots are the 5×6 lattices considered for $q = 2$ and the lower plot is the 6×6 lattice considered for $q = 4$.

per site $\Delta^{1/N}$, which is expected to be roughly independent of system size for large enough systems.

Let us first investigate the model with $q = 2$. The low energy part of the spectrum of the truncated Hamiltonian for different lattice sizes for square lattice, is shown in Fig. 7.4 and for a triangular lattice, is shown in Fig. 7.5, where we also provide the overlaps per site for the two lowest energy eigenstates. The truncation radius is $r = \sqrt{2}$ for the square lattice and $r = 1$ for the triangular lattice. Figure 7.6 gives a more detailed view with more eigenstates for the 4×4 lattices. The overlaps per site for the two lowest energy states are higher than 0.99 for all the cases, which shows that the wavefunctions are close to the analytical states derived from CFT. Already for the 4×4 lattice, the two lowest energy states are separated by a gap from the rest of the spectrum, which shows the twofold degeneracy on the torus, although not perfectly for this very small lattice size. The ratio $(E_2 - E_1)/(E_3 - E_1)$ is smaller for the 5×6 lattice than for the 4×4 lattice, and this suggests that a gap will be present in the thermodynamic limit.

As a further test of the topological nature of the ground states of the local models, we compute [112] the topological entanglement entropy γ for the states on the 5×6 lattice. The division of the lattice used for the computation is shown in Fig. 7.7. We do not expect to get very accurate results for this lattice size, since the condition that all regions should be much larger than the correlation length is not entirely fulfilled. The exact lattice Laughlin state at filling fraction $1/q$ has $\gamma = \ln(q)/2$, which is $\gamma \approx 0.347$ for $q = 2$. For the square lattice, we obtain $\gamma = 0.3216$ for the lowest energy eigenstate and $\gamma = 0.3839$ for the second lowest energy eigenstate, and for the triangular lattice, we obtain $\gamma = 0.3278$ for the lowest energy

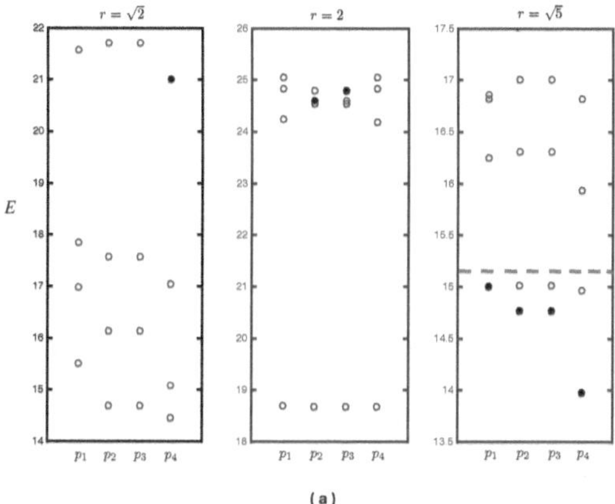

(a)

Fig. 7.8 Energy spectrum in the relevant momentum sectors $p_i = (p_x, p_y)$ for different truncation radii computed on a 6×6 square lattice on a torus. Here, p_x and p_y correspond to the momentum quantum numbers in the two periodic directions, respectively. The energy values painted black correspond to the eigenstates that have high overlap per site ($\Delta^{\frac{1}{N}} > 0.98$) with the exact analytical states. The green dashed line indicates the lowest energy present in the momentum sectors not shown.

eigenstate and $\gamma = 0.3412$ for the second lowest energy eigenstate. These values are close to the exact values and support the statement that the ground states are in the same topological phase as the half filled Laughlin states.

We next investigate the model with $q = 4$. We saw above that this case requires a larger radius of truncation. Testing the local models on small lattices with open boundaries for different truncation radii is possible, but on the torus the local regions will wrap around the torus and overlap with themselves if they are chosen too large. For a given lattice size, this puts a restriction on the largest truncation radius that can be studied numerically. We hence show the spectra for 6×6 square lattice in Fig. 7.8 and for a triangular lattice in Fig. 7.9. The plots show the energy spectrum in the momentum sectors where the analytic states of interest lie. We observe that the overlaps for the ground states are small for small truncation radii as expected. The trend in the plots, however, indicates that the states with significant overlap with the exact analytic states climb down the energy spectrum as the truncation radius is increased. For the square lattice and $r = \sqrt{5}$, there are four states with high overlap among the low energy states, but there is not a clear gap. We are not able numerically to study large enough systems to judge whether a gap will be present in the thermodynamic limit or the system will be in a different phase. It is interesting to note, however, that both for the

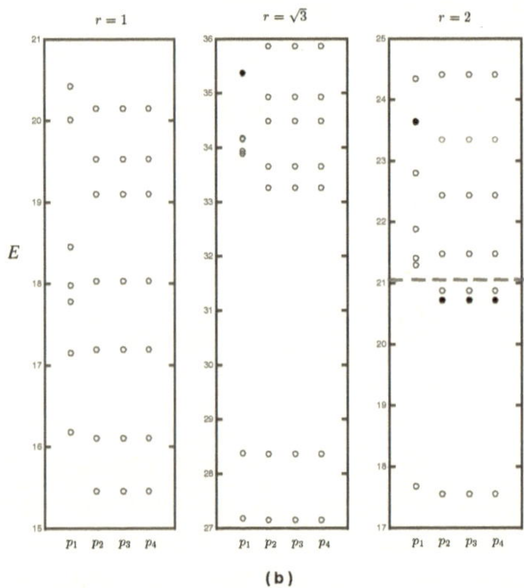

Fig. 7.9 Energy spectrum in the relevant momentum sectors $p_i = (p_x, p_y)$ for different truncation radii computed on a 6×6 triangular lattice on a torus. Here, p_x and p_y correspond to the momentum quantum numbers in the two periodic directions, respectively. The energy values painted black correspond to the eigenstates that have high overlap per site ($\Delta^{\frac{1}{N}} > 0.98$) with the exact analytical states. The green dashed line indicates the lowest energy present in the momentum sectors not shown.

4×4 lattice with open boundary conditions studied in Fig. 7.3 and for the 6×6 lattice with periodic boundary conditions, high overlap of the ground state is first observed for $r = \sqrt{5}$. This may suggest that $r = \sqrt{5}$ is also sufficient in the thermodynamic limits, but it would be necessary to investigate larger lattices to draw conclusions. For the local model at $r = \sqrt{5}$ on a 6×6 square lattice, we find $\gamma = 0.7088, 0.6231, 0.6894, 0.6724$ for the four lowest states that have significant overlap with the exact analytic states (the division used is shown in Fig. 7.7). This is close to the expected value $\gamma = \ln(4)/2 \approx 0.693$ and shows that the states with highest overlap with the exact analytic states are in the right topological phase.

For the triangular lattice, we also observe the trend that states with significant overlap are climbing down the spectrum as the truncation radius is increased. The results in Fig. 7.3 suggest that a truncation radius of at least $r = \sqrt{7}$ is needed to get high overlaps, which is consistent with the results in Fig. 7.9. For the 6×6 triangular lattice on the torus, we can, however, not study the case $r = \sqrt{7}$, since the largest allowed truncation radius is $r = 2$.

7.2   Local Hamiltonians for 1D critical models

Similar to 2D, one can also obtain 1D versions of the lattice Laughlin models using CFT methods [5]. The resulting models are critical and have interactions that decay as the inverse of the square of the distance $(1/r^2)$. One-dimensional quantum many-body models with nonlocal interactions of the kind $1/r^2$, as well as truncated versions of them, have gained considerable interest over the years [113, 114, 45, 46, 49, 55, 115, 86, 87, 5, 116, 117], exploring interesting aspects of integrable systems. A common feature of quantum $1/r^2$ systems is that the ground state can be exactly represented by a Jastrow-type wavefunction [113, 114, 45, 46, 49, 55, 115, 86, 87, 5, 116, 117], namely a product of two-body functions.

One purpose of constructing analytical models of quantum many-body systems with interesting properties is to use them as a guide to find ways to implement the physics experimentally. In many cases, it may not be possible to directly implement the analytical model itself, but with some suitable modifications one may be able to simulate a model displaying the same physics. Ultracold atoms in optical lattices provide an interesting and flexible setup for simulating quantum models, and there is a rapid development in the field [118]. As far as the models derived from conformal field theory are concerned, a challenging aspect for experimental implementation is that the models involve nonlocal interactions. This poses the question, whether the same physics can be realized in a model with only local interactions, and whether one can use the form of the exact model to predict a suitable choice of local interactions.

In this section, we investigate this question for the family of 1D models related to the Laughlin states. We do the investigation with exact diagonalization, and we find that it is possible to obtain local Hamiltonians, whose low-energy eigenstates have similar critical properties as the eigenstates of the exact Hamiltonians. Finally, we briefly discuss possibilities for implementing the local models in ultracold atoms in optical lattices.

7.2.1 Exact Models

Our starting point is a family of exact 1D models with non-local interactions that have been constructed using tools from a CFT in [5]. The models are defined on a 1D lattice with periodic boundary conditions, and it is convenient to think of the N lattice sites as sitting on a unit circle in the complex plane with the jth site at $z_j = e^{i2\pi j/N}$. Each site can be either empty or occupied by one particle, and n_j denotes the number of particles on site j. The members of the family are labeled by a positive integer q, and for q odd (even), the particles are fermions (hardcore bosons). The Hamiltonian takes the form

$$H_{1D} = \sum_{i \neq j} \left[(q-2)w_{ij} - w_{ij}^2 \right] d_i^\dagger d_j - \frac{1}{2}(q^2 - q) \sum_{i \neq j} w_{ij}^2 n_i n_j. \tag{7.10}$$

is an eigenstate of the Hamiltonian (7.10) with energy

$$E_0 = -\frac{(q-1)}{6q} N[3N + (q-8)], \tag{7.11}$$

The unique ground state $|\Psi_{\text{Exact}}\rangle$ in the subspace with N/q particles is given by Eq. (2.4) with $\eta = 1$ since we are in the strict lattice limit.

We comment that the Hamiltonian for $q = 2$ is closely related to the Haldane-Shastry spin Hamiltonian [45, 46]. This can be seen by introducing the transformation

$$S_i^+ = d_i^\dagger \quad S_i^- = d_i \quad S_i^z = d_i^\dagger d_i - 1/2, \tag{7.12}$$

where $S_i^\pm = S_i^x \pm i S_i^y$, and $\vec{S}_i = (S_i^x, S_i^y, S_i^z)$ is the spin operator acting on a two-level system. Inserting this into (7.10) and utilizing that the number of particles is fixed to $N/2$, we arrive at the Hamiltonian

$$H = \sum_{i \neq j} \frac{\vec{S}_i \cdot \vec{S}_j}{\tan^2\left(\frac{\pi(i-j)}{N}\right)} + \text{constant}. \tag{7.13}$$

This is the Haldane-Shastry Hamiltonian except for an additive and a multiplicative constant.

Monte Carlo simulations for systems with a few hundred sites have already shown that the exact model is critical with a ground state entanglement entropy that grows logarithmically with the size of the subsystem and ground state two-site correlation functions that decay as a power law [5]. In the following, we concentrate on the cases $q = 3$, 2, and 4. We do so, because we want to study both the fermionic and the hardcore bosonic case, and because the $q = 2$ model has an SU(2) invariant ground state, while the $q = 4$ and $q = 3$ models do not.

Let us investigate the behavior of the coefficients $C_1 = (q - 2)w_{ij} - w_{ij}^2$ of the hopping terms and $C_2 = -\frac{1}{2}(q^2 - q)w_{ij}^2$ of the density-density interaction terms of the exact Hamiltonian (7.10) with respect to distance. First we note that

$$w_{jk} = \frac{z_j + z_k}{z_j - z_k} = \frac{-i}{\tan[\frac{\pi}{N}(j - k)]}. \tag{7.14}$$

We have

$$w_{jk} \approx \frac{-iN}{\pi(j - k)} \qquad \text{for} \qquad \pi|j - k|/N \ll 1. \tag{7.15}$$

In this limit, the terms proportional to w_{ij}^2 are hence much larger than the term proportional to w_{ij}. It follows that C_1 and C_2 are of the same order, and they both decay as $|i - j|^{-2}$ for $\pi|i - j|/N \ll 1$. For long distances, where $|i - j| \approx N/2$, both C_1 and C_2 approach zero. The fact that the interaction strengths decay fast with distance suggest that it may be possible to truncate the nonlocal Hamiltonian to a local Hamiltonian without significantly altering the low-energy physics of the model.

7.2.2   Local models

Replacing the exact Hamiltonian with a local one is an advantage experimentally, both because it removes the need to engineer couplings between distant sites and because it reduces the number of terms in the Hamiltonian. Ultracold atoms in optical lattices provide an interesting framework for simulating quantum physics. By trapping ultracold atoms in a 1D optical lattice, one naturally gets a model with hopping between NN sites and on-site interactions [21]. The hardcore constraint can be implemented by ensuring that the on-site interaction is large enough. Density-density interactions between neighboring sites can be achieved through dipole-dipole interactions, see e.g. [119–121].

If one changes the geometry of the 1D lattice into a zigzag lattice, one can have couplings between both NN and NNN sites, as schematically depicted in Fig. 7.10. There are already more methods available to realize zigzag lattices with cold atoms, see e.g. [122–125]. Anisimovas *et. al.* have proposed a scheme for realizing a zigzag lattice [125] using ultracold bosons. In their study, the two legs of the ladder correspond to a synthetic dimension given by two spin states of the atoms, and the tunneling between them can be realized by a laser-assisted process. Subsequently, by employing a spin-dependent optical lattice with

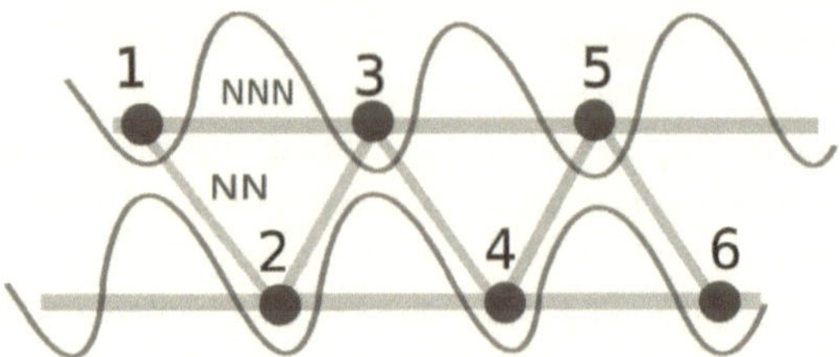

Fig. 7.10 Schematic diagram showing a zigzag lattice. The geometry allows for adjusting the strengths of the NNN terms relative to the strengths of the NN terms.

the site position depending on the internal atomic state, a zigzag ladder can be achieved. In another experimental proposal, Zhang *et. al.* have described a feasible method to achieve the zigzag lattice in one dimension [124]. They use a superlattice generated by commensurate wavelengths of light beams to realize tunable geometries including zigzag and sawtooth configurations. Also, in a theoretical study, Greschner *et. al.* have suggested that the zigzag chain may be formed by an incoherent superposition between a triangular lattice and a 1D superlattice [122].

In the following, we will therefore restrict ourselves to models with at most NNN couplings in the Hamiltonian. Specifically, we study four different models. The first one is the NN model defined by the Hamiltonian

$$H_{\mathrm{NN}} = \sum_{<i,j>} \left[(q-2)w_{ij} - w_{ij}^2\right] d_i^\dagger d_j - \frac{1}{2}(q^2 - q) \sum_{<i,j>} w_{ij}^2 n_i n_j. \qquad (7.16)$$

Here, the sum over i and j is restricted to NN terms as denoted by the symbol $< \ldots >$. (We include both, e.g., $(i,j) = (1,2)$ and $(i,j) = (2,1)$ in the sum, which ensures that the Hamiltonian is Hermitian.) For $q = 2$, H_{NN} reduces to the antiferromagnetic spin-$1/2$ XXX Heisenberg Hamiltonian except for an additive constant (this can be seen by applying the transformation in (7.12)).

The second model is the model obtained by truncating the full Hamiltonian to only include terms up to NNN distance, i.e.,

$$H_{\mathrm{NNN}} = \sum_{<<i,j>>} \left[(q-2)w_{ij} - w_{ij}^2\right] d_i^\dagger d_j - \frac{1}{2}(q^2 - q) \sum_{<<i,j>>} w_{ij}^2 n_i n_j, \qquad (7.17)$$

where $<< \ldots >>$ is the sum over NN and NNN terms.

Instead of simply cutting the Hamiltonian as done above, one could try to partially compensate for the removed terms by adjusting the relative strengths of the couplings. We

Table 7.1 Optimal values of U_1 and U_2.

q	U_1	U_2
2	1.00	1.00
3	1.86	0.68
4	5.36	0.60

therefore also consider the optimized NN model

$$H_{\mathrm{NN}}^{\mathrm{opt}} = \sum_{<i,j>} \left[(q-2)w_{ij} - w_{ij}^2\right] d_i^\dagger d_j - \frac{U_1}{2}(q^2 - q) \sum_{<i,j>} w_{ij}^2 n_i n_j, \qquad (7.18)$$

and the optimized NNN model

$$H_{\mathrm{NNN}}^{\mathrm{opt}} = \sum_{<<i,j>>} \left[(q-2)w_{ij} - w_{ij}^2\right] d_i^\dagger d_j - \frac{U_2}{2}(q^2 - q) \sum_{<<i,j>>} w_{ij}^2 n_i n_j. \qquad (7.19)$$

In these models, U_1 and U_2 are parameters that are chosen to maximize the overlap between the ground state of $H_{\mathrm{NN}}^{\mathrm{opt}}$ or $H_{\mathrm{NNN}}^{\mathrm{opt}}$ with the analytical state $|\psi_{\mathrm{Exact}}\rangle$ in Eq. (2.7).

We investigate how the overlap varies with U_1 and U_2 in figure 7.11. For a given q, the overlap value is seen to be maximized at a unique value of U_1 and U_2, independent of the system size N. The optimal values are given in table 7.1. We do not know why this is the case. For $q = 2$, the optimal value is unity, and the optimized models hence coincide with the models that are not optimized. It is interesting to note that the Hamiltonians for $q = 2$ and $U_1 = U_2 = 1$ are SU(2) invariant, as all terms can be written in terms of $\vec{S}_j \cdot \vec{S}_k$ (see (7.12)). For $q = 2$, the wavefunction Eq. (2.7) is also SU(2) invariant. It is hence natural that $U_1 = U_2 = 1$ is the optimal choice.

7.2.3 Properties of ground states of the local models

We now compute properties of the ground states of the local models to quantify how well the local models reproduce the physics of the exact models. The results are obtained using exact diagonalization, and we hence consider systems with $q = 2$, 3, and 4 with at most 32 lattice sites and N/q particles. We first show that overlaps per site higher than 0.999 can be obtained for all these cases with at most optimized NNN interactions. We then show that the behavior of important properties like entanglement entropy and two-site correlation functions is also reproduced well (possibly except for a bit of discrepancy for $q = 3$ and N even).

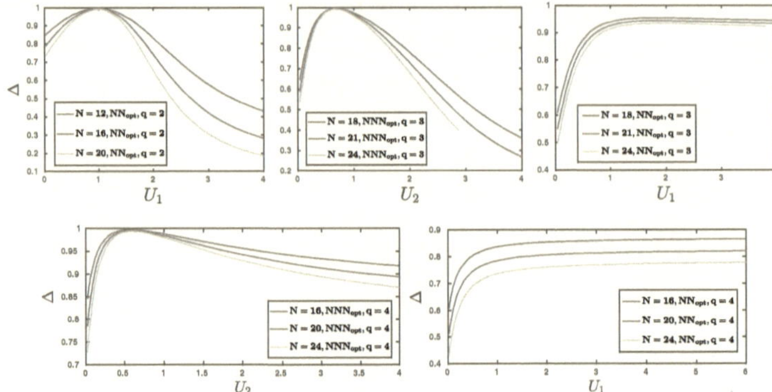

Fig. 7.11 Overlap value as a function of optimizing parameters U_1 and U_2 for $q = 3$, $q = 2$, and $q = 4$.

7.2.3.1 Overlap between ground state wave functions

We have calculated the overlap Δ between the ground state $|\Psi_{\text{Local}}\rangle$ of the local Hamiltonian (7.16), (7.17), (7.18) or (7.19) and the analytical wavefunction $|\Psi_{\text{Exact}}\rangle$ in Eq. (2.7). For a non-degenerate ground state, we define the overlap as

$$\Delta = |\langle\Psi_{\text{Local}}|\Psi_{\text{Exact}}\rangle|^2. \tag{7.20}$$

For large enough systems, the overlap is expected to decay exponentially with system size, even if the states deviate only a bit from each other, since the dimension of the Hilbert space increases exponentially in N. We therefore also consider the overlap per site $\Delta^{1/N}$, which is expected to remain roughly constant with system size for large enough N. We find that the ground state is non-degenerate for all the system sizes we consider.

The overlap for the fermionic case ($q = 3$) is given in table D.1 in the appendix. For a complete analysis, these Δ values are explicitly calculated with and without optimization for different system sizes. The overlaps are higher for the NNN Hamiltonian than for the NN Hamiltonian, and optimization also improves the overlaps. The overlap decreases with system size, but the overlap per site is almost constant. The overlap per site is already around 0.996 for only NN interactions without optimization, and for the NNN Hamiltonian with optimzation it is around 0.9997.

Next, we report the overlap values for half-filling ($q = 2$) and quarter-filling ($q = 4$) of the lattice with hardcore bosons. These results are shown in tables D.2 and D.3 in the appendix,

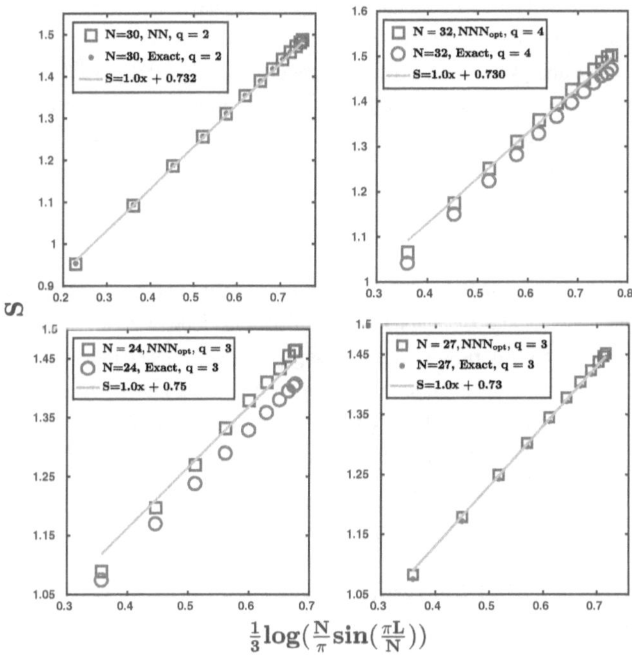

Fig. 7.12 Entanglement entropy S as a function of subsystem size L for $q = 2$, $q = 4$ and $q = 3$. The numerical results for S are computed using the ground states of H_{NN} for $q = 2$ and $H_{\mathrm{NNN}}^{\mathrm{opt}}$ for $q = 3$ and $q = 4$. The different plots are for different q values as specified in the legends. The equations for the straight lines with slope 1 are also shown in the legends.

respectively. For $q = 2$, the NN model is enough to obtain overlaps per site around 0.9995, and these overlaps are slightly improved by considering the NNN model.

The scenario is different for $q = 4$. In this case, the overlap per site is around 0.987 for the NN model. Adding NNN interactions increases the overlap per site to around 0.999. The optimized NN model produces lower overlaps than the NNN model. Optimizing the NNN model gives overlaps slightly better than for the NNN model. For $q = 4$, the average distance between particles is four sites, which could be part of the explanation why NNN interactions are more important in this case than for $q = 2$.

From the above analysis of the overlap, we conclude that for the $q = 2$ case, it is sufficient to consider the local Hamiltonian with only NN interactions to study the ground state properties, whereas for the $q = 3$ and $q = 4$, the local NNN Hamiltonian with optimization is an appropriate choice.

7.2.3.2 Entanglement Entropy

We calculate the von Neumann entanglement entropy defined in Eq. (4.13) for our 1D system. We consider subsystems that consist of L consecutive sites. Due to the translational invariance of the models we are considering, all such entanglement entropies can be expressed in terms of the entanglement entropies obtained when region A is site number 1 to site number L with $L \in \{1, 2, \ldots, N/2\}$ for N even and $L \in \{1, 2, \ldots, (N+1)/2\}$ for N odd.

The von Neumann entanglement entropy of a critical system in 1D is given by [126]

$$S = \frac{c}{3} \log \left[\frac{N}{\pi} \sin \left(\frac{\pi L}{N} \right) \right] + \text{constant}, \tag{7.21}$$

when L is large compared to 1. The central charge c was found to be $c = 1$ for the exact model in [5]. Plotting S as a function of $(1/3) \log[(N/\pi) \sin(\pi L/N)]$, we should hence compare to a line of slope 1.

Plots of the entanglement entropy for $q = 3$, 2, and 4 and different system sizes are shown in Fig. 7.12. For $q = 2$, the data for the NN model and the exact model are seen to fit very well and also fit well with a line of slope 1. For $q = 3$, there is a good match for N odd, while there is a bit of deviation for N even. For $q = 4$, the data for the local model are displaced by a constant, but the slope for L not too small fits well with 1. The central charge is hence the same as in the exact model.

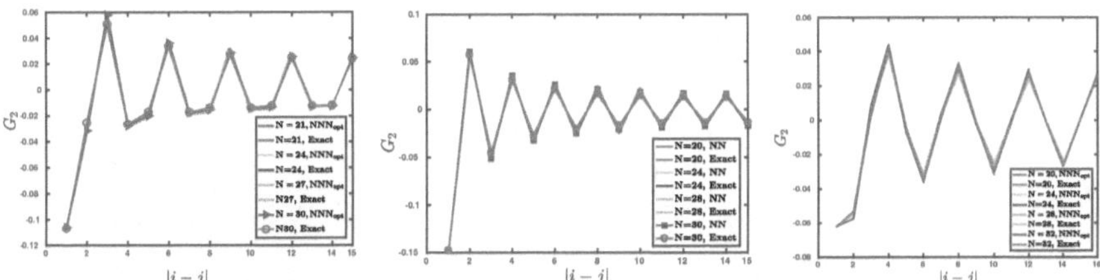

Fig. 7.13 Correlation function G_2 for $q = 3$ (left), $q = 2$ (middle), and $q = 4$ (right). Results are shown both for the wavefunction $|\Psi_{\text{Exact}}\rangle$ and for the ground state of the local Hamiltonian ($H_{\text{NNN}}^{\text{opt}}$ for $q = 3$ and $q = 4$ and H_{NN} for $q = 2$).

7.2.3.3 Correlation function

Correlation functions are also an important characteristic property for determining the physics of a system. Here, we compute the two-site correlation function defined as

$$G_2 \;=\; \langle\Psi|n_i n_j|\Psi\rangle - \langle\Psi|n_i|\Psi\rangle\langle\Psi|n_j|\Psi\rangle, \tag{7.22}$$

where $|\Psi\rangle$ is the state of the system. In our case, $\langle\Psi|n_j|\Psi\rangle = 1/q$ for all j. In the exact model, G_2 decays with the distance between the points to the power $-2/q$ and also shows oscillations with period q. For the local models, it is difficult to determine the precise behavior of the decay, since we cannot go to sufficiently large system sizes. Instead we simply compare the values of G_2 for the local and the exact models. Plots of G_2 are shown for $q = 3$, 2, and 4 and different system sizes in Fig. 7.13. In all cases, there is a good agreement between the local and the exact models.

7.2.4 Low-lying excited states

Achieving a good overlap for the ground state wave functions is a good starting point, but if the system is at nonzero temperature or we are interested in dynamics, it is also important that the low-lying excited states are not affected significantly in going from the exact to the local Hamiltonian, and we now investigate this question.

The energies of the seven lowest states are plotted in Figs. 7.14, 7.15, and 7.16 for $q = 3$, 2, and 4, respectively, for both a local model and the exact model. We have added a constant to the energies and divided by a constant factor to fix the ground state energy to zero and the energy of the first excited state to one. For $q = 3$, it is seen that the energies of the lowest excited states of the local model fit accurately with the energies of the corresponding states for the exact model, in particular for the first four excited states. For $q = 2$, the energies of

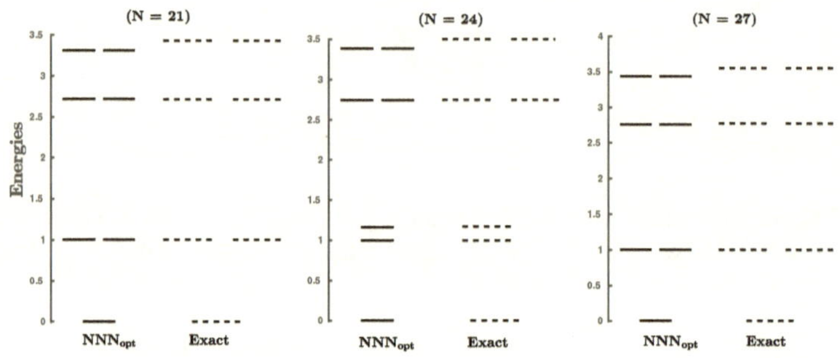

Fig. 7.14 Energies of the 7 lowest states of the NNN_{opt} model (7.19) (shown as solid lines) and of the exact model (7.10) (shown as dashed lines) for $q = 3$ (the ground state energy has been set to zero, and the energies have been scaled by a constant factor such that the first excited state is at energy 1). The different plots are for 21, 24, and 27 sites, respectively, as indicated.

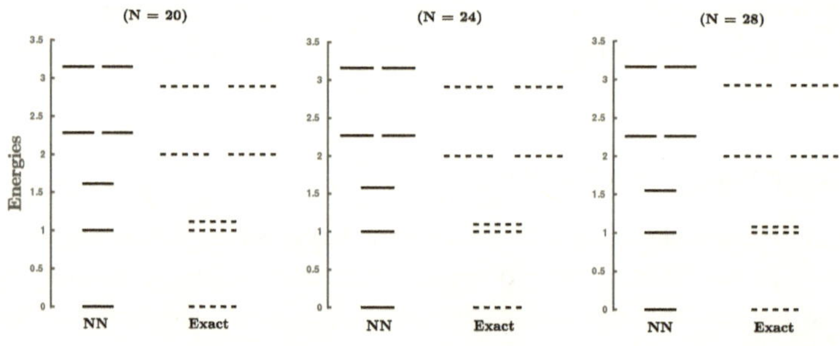

Fig. 7.15 Energies of the 7 lowest states of the NN model (7.16) (shown as solid lines) and of the exact model (7.10) (shown as dashed lines) for $q = 2$ (the ground state energy has been set to zero, and the energies have been scaled by a constant factor such that the first excited state is at energy 1). The different plots are for 20, 24, and 28 sites, respectively, as indicated.

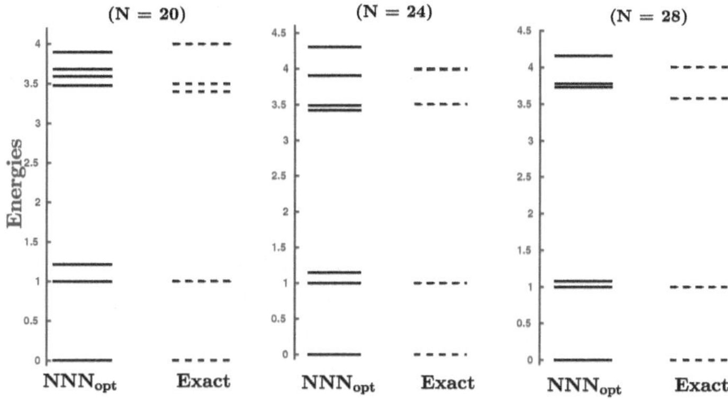

Fig. 7.16 Energies of the 7 lowest states of the $\mathrm{NNN}_{\mathrm{opt}}$ model (7.19) (shown as solid lines) and of the exact model (7.10) (shown as dashed lines) for $q = 4$ (the ground state energy has been set to zero, and the energies have been scaled by a constant factor such that the first excited state is at energy 1). The different plots are for 20, 24, and 28 sites, respectively, as indicated.

the lowest excited states in the local model are generally shifted compared to the exact model, but the degeneracies remain the same. In particular, it is seen that the energy difference between the first and the second excited state is much larger for the local model than for the exact model. For $q = 4$, it is again the case that the energies of the lowest excited states in the local model are shifted compared to the exact model, but the degeneracies remain the same. It is interesting to note that the plots for a given value of q are almost independent of N for $q = 2$ and $q = 3$, but not for $q = 4$. This shows that finite size effects play a role for $q = 4$. Our exact diagonalization codes do, however, not allow us to go to large enough systems to eliminate these effects.

We next compute overlaps to investigate how well the wavefunctions of the low-energy excited states of the local models match with the corresponding eigenstates of the exact models. We find that the overlap per site is higher than 0.998 in all the considered cases, which shows that there is an excellent agreement. Details of the results can be found in tables D.4, D.5, and D.6 in the appendix. The fact that the overlaps are nonzero also shows that there is agreement between all quantum numbers of the states of the local Hamiltonian and the states of the exact Hamiltonian.

The conclusion of this subsection is that the low-energy excited state wavefunctions of the local models are close to the corresponding wavefunctions of the exact models, but the energies may be shifted. If we prepare the system in a given low-energy excited state, we hence expect the physics to be practically the same for the local and the exact model. If we instead consider a superposition of different excited states, the relative phases between the terms in the superposition may have different time variation.

7.3 Conclusion

In the first part of this chapter, we discussed a general procedure to truncate lattice FQH Hamiltonians derived from CFT null fields. Each of the null fields leads to an operator Λ_i that annihilates the state, and the approach is to truncate this operator to a local form instead of truncating the Hamiltonian itself. Truncating the Λ_i operator has a number of advantages compared to truncating the Hamiltonian directly. First, the terms in the operator depend only on the relative coordinates of the involved sites, and not on the rest of the lattice. The local terms are hence unaltered, when we approach the thermodynamic limit or in other ways modify the rest of the lattice. On the contrary, some of the terms in the Hamiltonian depend on the whole lattice, and the local terms resulting from truncating the Hamiltonian directly will depend on the choice and size of the whole lattice. Second, it is clear how to construct models with open or periodic boundary conditions as desired, while truncating the Hamiltonian directly only gives models with open boundary conditions, unless one manually changes the local terms and optimizes the coupling strengths numerically. The fact that numerical optimization is not needed means that we can construct the local Hamiltonian even if the local regions are not small compared to the lattice sizes that can be investigated with exact diagonalization.

We have applied the truncation procedure to Hamiltonians with half and quarter filled Laughlin type ground states on square and triangular lattices on the plane and on the torus. For the $q = 2$ local model, we find that a truncation radius of $r = \sqrt{2}$ on the square lattice and $r = 1$ on the triangular lattice is enough to obtain overlaps per site with the exact states that are larger than 0.99 for all the investigated cases, and to obtain an approximate twofold degeneracy on the torus. This suggests that the truncated model is in the same topological phase as the exact model. We also find that the topological entanglement entropy is close to the expected value. For the $q = 4$ local model, we infer first from computations assuming open boundary conditions that a larger truncation radius is needed for the ground state of the local model to be close to the exact analytical state. We see a similar trend on the torus. Computations for the square lattice for open boundaries and on the torus seem to suggest that a truncation radius $r = \sqrt{5}$ is needed to stabilize the corresponding $q = 4$ Laughlin state. The systems we can investigate with exact diagonalization are, however, too small to judge if the system is, indeed, topological in the thermodynamic limit. We have also checked that the states have the right topological entanglement entropy. For the case of the triangular lattice, the computations for open boundary conditions indicate that at least a truncation radius of $r = \sqrt{7}$ is needed for the ground state of the local model to be sufficiently close to the exact analytical state. However, numerical restrictions do not allow us to test this on larger lattice sizes on the torus.

It is worth mentioning that an SU(2) invariant local Hamiltonian was constructed for the $q = 2$ state on the square lattice and the kagome lattice in Ref. [7] by truncating the Hamiltonian directly and numerically optimizing the coefficients. The same procedure does, however, fail to produce a simple Hamiltonian on the triangular lattice. In contrast, the approach of truncating the Λ_i operator gives good overlap values on both the square and the triangular lattice without optimization. Also, for the FQH Hamiltonians considered in this chapter, there is no SU(2) symmetry to simplify the problem. Hence, this way of truncating the Hamiltonian should work generally for a larger class of model Hamiltonians derived from CFT.

In the second part of the chapter, we started from a class of one-dimensional models derived from conformal field theory and used exact diagonalization to investigate to what extent the same physics is realized by different models containing only NN and possibly NNN couplings. For the $q = 2$ model we found that already the NN model, which is equivalent to the antiferromagnetic XXX Heisenberg model, gives a good description. For $q = 3$ and $q = 4$, an optimized model with both NN and NNN couplings is preferred. With these models we find that the overlap per site between the ground states of the local models and the ground states of the exact models is larger than 0.999 for the considered system sizes. The overlaps per site for the low-lying excited state are larger than 0.998. We also find that the ground state entanglement entropy and the correlation functions are reproduced well by the local models, except that there is a bit of discrepancy for $q = 3$ for N even for the entanglement entropy. The low-energy spectra are modified to some extent in the local models compared to the exact models. For $q = 4$, the modification varies with system size, so finite size effects play a role for the considered system sizes. The high agreement between the ground state properties and the excited states suggest that most of the physics will remain the same in the local models. This paves the way to experimentally realize the physics in optical lattices using ultracold atoms, since the local models are simpler to realize experimentally than the corresponding exact models derived from conformal field theory. The ingredients needed to realize the different terms in the Hamiltonian are already available experimentally.

Hence, the main observation from this chapter is that, conformal field theory can not just be used to obtain exact, nonlocal models, but is also a powerful tool to obtain interesting, local models.

Chapter 8

Braiding anyons effectively on a small lattice

As we discussed in Chapter 2, lattice has several advantages over the continuum in realizing FQH physics. One of the important reasons to study lattice FQH models is to create anyons and perform braiding in a controlled fashion. We remarked that potential traps can be included at specific lattice positions to pin the fractional defects which turn out be anyons in the topological phase. Then, the properties of the anyons may be investigated by studying the ground state of the many-body Hamiltonian. Braiding of anyons on lattice systems have been numerically studied on torus geometries where a relatively larger braiding path can be chosen compared to open boundaries. [40, 42, 44].

Although the above description allows for the existence of anyons, the charge of these defects are typically spread out over few lattice spacings. However, to get the topological braiding properties, the anyons need to be well separated and hence the lattice system needs to have a certain size. The computational resources needed to study a many-body quantum system exponentially grows with system size and hence there is an obvious limitation for studying braiding statistics for the anyons. Further, experimentally, it is generally challenging to keep quantum coherences in large systems and hence dealing with small system sizes is an advantage for ultracold atoms in optical lattices [41]. While numerically it may be advantageous to braid anyons on a torus geometry, care must be taken such that the anyons are sufficiently in the bulk and not overlap with each other while they are braided on a lattice with open boundaries. The challenge is hence to localize the anyons sufficiently while we braid them on a lattice with a smaller area.

In this chapter, we discuss how we address the above challenge by showing that the lattice anyons may be squeezed so that they are localised on fewer sites which facilitate the braiding even on small sized lattices. We demonstrate this first in a toy model by constructing lattice wavefunctions for squeezed anyons and derive the braiding statistics under some assumptions. We then consider the Kapit-Mueller model and an interacting Hofstadter model as examples of lattice fractional quantum Hall models. We choose open boundary conditions, since this is the most relevant case for experiments. The squeezing of the anyons is done with a position dependent potential. For the system size considered, the anyons overlap significantly when they are created from a local potential, and the phase acquired when two anyons are exchanged differs from the ideal value predicted for well separated anyons. Using the optimized potential removes the overlap of the charge distributions of the anyons thar now squeezed. Next, we braid these squeezed anyons first in the adiabatic limit by moving them like snakes and the braiding phase we get is significantly closer to the ideal value. The modifications done to shape the anyons could affect the size of the gap and the ability to couple to the excited states, and we therefore also compute the time needed for doing the exchange process in order to be close to the adiabatic limit. We also show numerically that if the potential is modified slightly away from the optimal choice, the phase acquired when exchanging two anyons in the considered finite size system remains practically the same and hence robust to small perturbations.

This chapter is based on the following reference [127].

- " *Squeezing anyons for braiding on small lattices* ", **N. S. Srivatsa**[*†], Xikun Li[*] & A. E. B. Nielsen, **Phys. Rev. Research 3, 033044 (2021)** [* authors contributed equally to this work].

8.1 Shaping anyons in a toy model

Topological properties are robust against local deformations, as long as the deformations are not so large that they bring the system out of the topological phase or allow the topological quantity in question to switch to one of the other allowed values. We therefore expect that it should be possible to shape the anyons to some extent, while keeping the braiding properties unaltered. We start out by showing this explicitly for a simple model, namely a modified Laughlin trial state on a lattice. This system can be analyzed using a combination of analytical observations and Monte Carlo simulations, and this allows us to study large systems with well-separated anyons.

We first consider the case of creation of anyons without shaping discussed in Chapter 2. As an example, we plot $\rho(z_j)$ defined in Eq. (2.15), for the state $|\psi_{2,2}\rangle$ given by Eq. (2.13), with two quasiholes of charge $1/2$ in Fig. 8.1(a). Each of the quasiholes occupies a roughly circular region and spreads over about 16 lattice sites. We obtain $-\sum_{j\in B}\rho(z_j) \approx 0.5$ as expected, where B is the set of the labels of the 16 sites inside one of the dashed circles.

Figure 8.1(b) shows $\rho(z_j)$ along a line parallel to the real axis that goes through one of the quasiholes.

The idea is to shape the quasiholes by splitting the factors $(w_j - z_k)$ appearing in the wavefunction into several pieces. The splitting gives us additional parameters that we can tune to obtain the desired shape. We first study how the splitting affects the density difference, and in the next section we will show under which conditions the shaped objects have the same braiding statistics as the original anyons. Specifically, we define the L weights g_j^h that are real numbers and the L coordinates w_j^h that are complex numbers. The weights fulfil the constraint $\sum_{h=1}^{L} g_k^h = p_k$. The resulting state takes the form

$$\psi_{q,S}(n_1, n_2, \ldots, n_N) = C^{-1} \delta_n \prod_{h,j,k} (w_j^h - z_k)^{g_j^h n_k} \times \prod_{i<j} (z_i - z_j)^{q n_i n_j} \prod_{i \neq j} (z_i - z_j)^{-n_i}. \quad (8.1)$$

The state (2.13) is the special case with $L = 1$.

We show one example in Fig. 8.1(c), where we plot the density profile for the state (8.1) with $q = 2$, $S = 2$, $L = 3$, and $g_j^h = 1/3$ for all h and j. The coordinates w_j^h are chosen as illustrated with the pluses in the figure. In this case, we find that $\rho(z_j)$ is nonzero in two regions that are narrower in one direction and about the same size in the other direction compared to the quasihole shapes in Fig. 8.1(a). The result is hence squeezing. We find that $-\sum_{j \in B} \rho(z_j) \approx 0.5$, where the set B consists of the labels of all the lattice sites inside one of the dashed ellipses. The squeezing is also seen in Fig. 8.1(d), where we show $\rho(z_j)$ along a line parallel to the real axis. By making different choices of g_j^h and w_j^h, we can obtain different shapes.

8.1.1 Braiding statistics

We now investigate the braiding properties of the shaped objects. The computations done here generalize the computations for $L = 1$ in [6, 128]. When we take the rth shaped object around a closed path, the wavefunction transforms as $|\psi_{q,S}\rangle \rightarrow M e^{i\theta_r} |\psi_{q,S}\rangle$, where M is the monodromy, which is determined by the analytical continuation properties of the wavefunction, and

$$\theta_r = i \sum_{h=1}^{L} \oint_{c_h} \langle \psi | \frac{\partial \psi}{\partial w_r^h} \rangle dw_r^h + \text{c.c.} \quad (8.2)$$

$$= \frac{i}{2} \sum_{h=1}^{L} \sum_{k=1}^{N} g_r^h \oint_{c_h} \frac{\langle n_k \rangle}{w_r^h - z_k} dw_r^h + \text{c.c.}$$

is the Berry phase. Here, we denote the path that w_r^h follows by c_h.

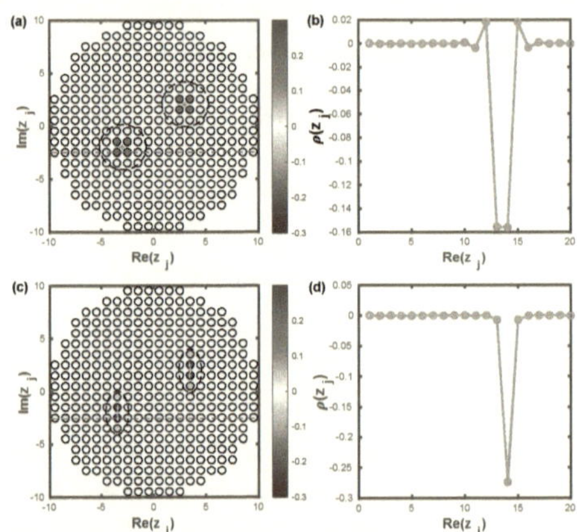

Fig. 8.1 (a) Density difference $\rho(z_j)$ [see Eq. (2.15)] of two quasiholes with charge 1/2 in the state (2.13). The quasihole positions w_k are depicted by red plus symbols. The charge has a spread of roughly 16 lattice sites (the sites that fall within the dashed circles). (b) Density difference along the row marked by the purple line in (a). (c) Density difference of squeezed quasiholes of charge 1/2 constructed by using three weights with $g_1^h = g_2^h = 1/3$ (at the positions indicated by the green and yellow plus symbols) using Eq. (8.1). The charge has a spread of roughly 8 lattice sites (the sites that fall within the dashed ellipses). (d) The density difference on the lattice sites along the row marked by a purple line in (c) also shows the squeezing. In both cases, $q = 2$, the number of lattice sites is $N = 316$, and we choose a square lattice with a roughly circular edge to mimic a quantum Hall droplet.

The phase θ_r contains contributions both from statistical phases due to braiding and from the Aharonov-Bohm phase acquired from the background magnetic flux encircled by the quasihole. Here, we want to isolate the statistical phase when the rth anyon encircles the vth anyon. The quantity of interest is hence

$$\theta_{\mathrm{in}} - \theta_{\mathrm{out}} = \frac{i}{2} \sum_{h=1}^{L} \sum_{k=1}^{N} g_r^h \oint_{c_h} \frac{\langle n_k \rangle_{\mathrm{in}} - \langle n_k \rangle_{\mathrm{out}}}{w_r^h - z_k} dw_r^h + \mathrm{c.c.}, \tag{8.3}$$

where θ_{in} ($\langle n_k \rangle_{\mathrm{in}}$) is the Berry phase (particle density) when the vth anyon is well inside all the c_h, and θ_{out} ($\langle n_k \rangle_{\mathrm{out}}$) is the Berry phase (particle density) when the vth anyon is well outside all the c_h. All other anyons in the system stay at fixed positions to avoid additional contributions to the Berry phase.

The density difference $\langle n_k \rangle_{\mathrm{in}} - \langle n_k \rangle_{\mathrm{out}}$ is nonzero only in the proximity of the two possible positions of the vth anyon and is hence independent of w_r^h. This facilitates to move the density difference outside the integral. The remaining integral $\oint_{c_h} \frac{1}{w_r^h - z_k} dw_r^h$ is $2\pi i$ whenever z_k is inside c_h. Due to the assumption that the vth anyon is well inside or well outside c_h, it follows that $-\sum_{k \text{ inside } c_h} (\langle n_k \rangle_{\mathrm{in}} - \langle n_k \rangle_{\mathrm{out}}) = p_v/q$ is the charge of the vth anyon. Utilizing further that $\sum_h g_r^h = p_r$, we conclude

$$\theta_{\mathrm{in}} - \theta_{\mathrm{out}} = 2\pi p_r p_v / q. \tag{8.4}$$

The result is hence independent of the number of weights L as long as the anyons are well separated and all the weights belonging to one anyon are moved around all the weights of the other anyon.

Looking at the monodromy, a difficulty immediately arises. In the normal Laughlin state, there is a trivial contribution to the monodromy, when an anyon encircles a particle, but this is not necessarily the case here. If the weight at w_r^h encircles the lattice site at z_j, the contribution to the monodromy is $e^{2\pi i g_r^h n_j}$, and this is nontrivial when g_r^h is not an integer. In order to get Laughlin type physics, we hence need that all such factors combine to a trivial phase factor. This can be achieved by putting a restriction on how the anyons are moved. Specifically, if we require that all c_h encircle the same set of lattice sites, then we get the contribution $e^{2\pi i \sum_h g_r^h n_j} = 1$ for each lattice site inside the paths.

The conclusion is hence that we obtain the same braiding properties as for the normal Laughlin state as long as all the weights of an anyon encircle all the weights of another anyon and the closed paths followed by the weights belonging to an anyon all encircle the same lattice sites. This still leaves a considerable amount of freedom to shape the anyons.

8.2 Squeezing and braiding anyons in the Kapit-Mueller model and in an interacting Hofstadter model

We next investigate anyons in the Kapit-Mueller model [129] and in an interacting Hofstadter model [17, 16]. We choose open boundary conditions, since this is the most relevant case for experiments. We first show that one can squeeze the anyons by adding an optimized potential, and in this way it is possible to avoid significant overlap between the charge distributions of the anyons even for the small system sizes considered. We then braid the squeezed anyons and find that the Berry phase is closer to the ideal value than it is for the case without squeezing. We also demonstrate robustness with respect to small errors in the optimized potentials and estimate the time needed to reach the adiabatic limit.

8.2.1 Model

We consider hardcore bosons on a two-dimensional square lattice with N sites and open boundary conditions. The hardcore bosons are allowed to hop between lattice sites, and the phases of the hopping terms correspond to a uniform magnetic field perpendicular to the plane. We study two models. The first one is the Kapit-Mueller model at half filling [129] with Hamiltonian

$$H_0 = \sum_{j \neq k} t(z_j, z_k) a_j^\dagger a_k + \text{H.c.}, \tag{8.5}$$

where a_j is the operator that annihilates a hardcore boson on the lattice site at the position z_j. Note that H_0 conserves the number of particles $\sum_i n_i$, where $n_i = a_i^\dagger a_i$ is the number operator acting on site i. We choose the lattice spacing to be unity, and take the origin of the coordinate system to be the center of the lattice. The coefficient for the hopping from the site at z_k to the site at z_j is then

$$t(z_j, z_k) = (-1)^{\text{Re}(z_j - z_k) + \text{Im}(z_j - z_k) + \text{Re}(z_j - z_k)\,\text{Im}(z_j - z_k)}$$
$$\times t_0 e^{-(\pi/2)(1-\phi)|z_j - z_k|^2} e^{-i\pi\phi[\text{Re}(z_j) + \text{Re}(z_k)]\,\text{Im}(z_j - z_k)}, \tag{8.6}$$

where ϕ is the number of magnetic flux units per site. We choose $t_0 = 1$ as the energy unit. When there are no anyons in the system, half filling means that the number of particles $\sum_i n_i$ is half the number of flux units $N\phi$. When pinning potentials are inserted to trap S anyons, we should instead choose the number of particles such that $\sum_i n_i + S/2 = N\phi/2$. Note that $t(z_j, z_k)$ decays as a Gaussian with distance between the two sites. If we only allow hopping between nearest neighbor sites, i.e. $\sum_{j\neq k} \mapsto \sum_{\langle j,k\rangle}$, then the resulting Hamiltonian is the interacting Hofstadter model [17, 16]. This is the second model we study.

Both models are in the same topological phase as the bosonic lattice Laughlin state with $q = 2$ for appropriate choices of the parameters. For the numerical computations below, we take an $N = 6 \times 6$ lattice with magnetic flux density $\phi = 1/6$, and we have either three particles and no quasiholes or two particles and two quasiholes in the system. Below, we describe how we create quasiholes and braid them.

8.2.2 Anyon squeezing and optimization

In Ref. [40], Kapit *et al.* considered a scenario in which quasiholes are trapped by local potentials. One can pin quasiholes at the sites a and b by adding the term $H_p = V n_a + V n_b$ to the Hamiltonian described in Eq. (8.5), where V is a positive potential strong enough to trap quasiholes. Let $\langle n_i\rangle_{H_0+H_p}$ be the expectation value of n_i in the ground state of the Hamiltonian with trapping potentials $H_0 + H_p$ when there are two particles and two quasiholes in the system, and let $\langle n_i\rangle_{H_0}$ be the expectation value of n_i in the ground state of H_0 when there are three particles and no quasiholes in the system. The density difference

$$\rho(z_i) = \langle n_i\rangle_{H_0+H_p} - \langle n_i\rangle_{H_0} \tag{8.7}$$

quantifies how much the presence of the anyons alter the density, and $-\rho(z_i)$ is the charge distribution of the anyons. In the ideal case of well separated anyons, the density difference is zero everywhere, except in the vicinity of the anyons, and each anyon appears as one half particle missing on average in a local region. In Fig. 8.2(b), we show the density difference when the trapping potentials are located at the sites $(2,2)$ and $(5,5)$. Notice that the overlap between the two anyons is large compared to the absolute value of the charge at the location of the trapping potentials. One could alternatively think of having several potentials in a row as in Fig. 8.2(c), but also in this case there is a large overlap between the anyons.

Here, we show that one can avoid the overlap between the charge distributions of the two anyons by choosing the potentials appropriately, and we use optimal control to find the appropriate potentials. We apply the method to both the Kapit-Mueller and the interacting Hofstadter models. For each of the anyons we choose four consecutive sites on which we want to localize the anyon. We choose the potentials on these sites to be $(1-\lambda)V_0$, V_0, V_0, and λV_0, respectively, where $\lambda \in [0,1]$. Later, we shall use the parameter λ to move the anyons during the braiding operation. In addition, we introduce auxiliary potentials V_i^{aux} on all the other sites i_{aux} to achieve screening. By tuning the values of V_i^{aux}, using optimal control theory, we are able to localize each of the anyons to very high precision on the chosen

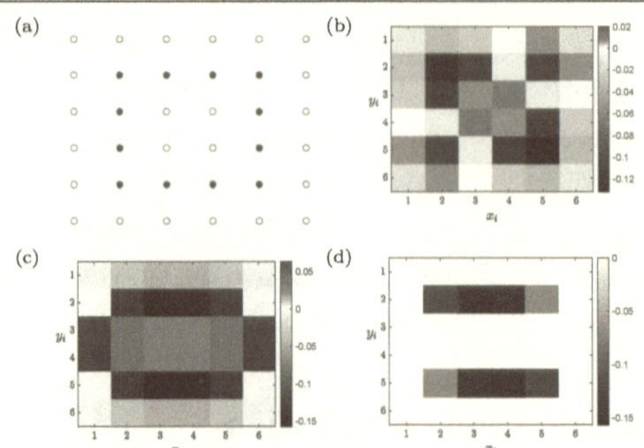

Fig. 8.2 (a) The path we choose to exchange the two quasiholes in the clockwise direction.
(b) The density difference for the case of two trapping potentials at the sites $(2,2)$ and $(5,5)$.
[We here label the lattice sites by (x_i, y_i), where $z_i = x_i - 3.5 - i(y_i - 3.5)$.] It is seen that
the two quasiholes trapped by the potentials overlap. (c) The density difference for the case
of trapping potentials on all the sites with $x_i = 2,3,4,5$ and $y_i = 2,5$. In this case, there is
again a significant overlap between the two anyons trapped by the potentials. (d) The density
difference for the case, where we optimize the potential to localize the quasiholes on the
eight sites. The sum of the absolute values of the charge on all the other sites is negligible
with $F \sim O(10^{-15})$.

sites. We study the ground state of the Hamiltonian $H = H_0 + H_p$, where H_0 is given by Eq.
(8.5) and the term $H_p = \sum_j V_j a_j^\dagger a_j$ includes the potential on all the sites.

We choose the fitness function

$$F(V_0, V_i^{\mathrm{aux}}) = \sum_{i \in i_{\mathrm{aux}}} |\rho(z_i)| \tag{8.8}$$

for the optimization to be the sum of the absolute value of the density difference over all the
auxiliary sites. Note that F is zero, when the anyons are perfectly localized on the chosen
sites. We adopt the covariance matrix adaptation evolution strategy (CMA-ES) algorithm to
minimize F. The CMA-ES belongs to the class of evolutionary algorithms and is a stochastic,
derivative-free algorithm for global optimization. It is fast, robust and one of the most popular
global optimization algorithms. See Ref. [130] for further details. We restrict the values
of the auxiliary potentials to be within the interval $V_i^{\mathrm{aux}} \in [-\varepsilon, \varepsilon]$, and their values within
this window are determined by the optimization algorithm. The hyper parameters (V_0, ε) are
chosen empirically, but we observe that the optimization is not sensitive to the values of V_0
and ε, as long as V_0 and ε are larger than certain threshold values. After the optimization
the fitness function is very small with $F \sim O(10^{-15})$, and each anyon is very well localized

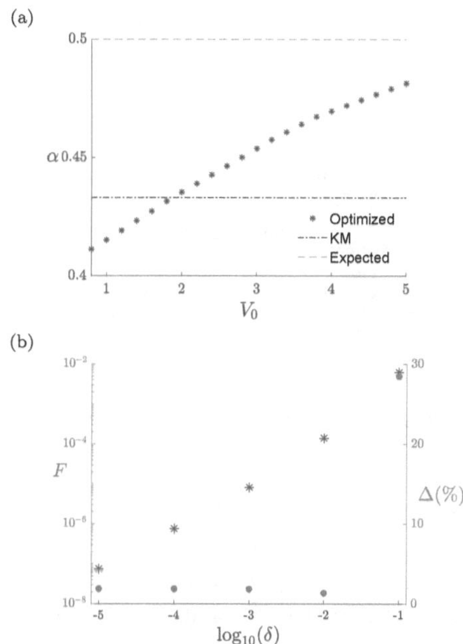

Fig. 8.3 (a) The Berry phase in units of π versus the strength of the trapping potentials V_0. The blue stars are for the squeezed anyons in the Kapit-Mueller model. The red dash-dotted line denotes the expected value of the Berry phase for well separated anyons. The black dashed line marks the best value obtained in Table 8.1 for the Kapit-Mueller model without optimization. (b) The robustness for the case with optimization against the strength of an error introduced in the potentials (see text for details). We display the values of the fitness function F (blue stars) in the left axis, as well as the relative error Δ (red dots, displayed as percentages) in the right axis, once the error δ is introduced.

within the desired four sites. See Fig. 8.2(d) for an illustration. In general, the optimized values of V_i^{aux} are smaller than V_0.

8.2.3 Adiabatic exchange

To exchange the anyons, we need the potentials to vary in time. For the case without optimization considered in [40], the potentials were moved between two sites by linearly increasing the strength at one site and linearly decreasing the strength at the other site. In our case, we move the chain of trapping potentials forward by one site by linearly increasing λ from 0 to 1. We move both the anyons simultaneously in the clockwise direction. We

Without optimization (KM/IH)			
N_{steps}	$V_0 = 1$	$V_0 = 10$	$V_0 = 10^2$	$V_0 = 10^3$
5	0.2805/0.3173	0.4088/0.4476	0.4307/0.4685	**0.4329/0.4706**
10	0.2815/0.3186	0.4098/0.4478	0.4308/0.4684	**0.4329/0.4706**
10^2	0.2819/0.3190	0.4096/0.4459	0.4304/0.4661	0.4329/0.4703
10^3	0.2819/0.3191	0.4095/0.4458	0.4296/0.4636	0.4324/0.4678

With optimization		
N_{steps}	$V_0 = 5$	$V_0 = 7$
4	**0.4873** (KM)	**0.5062** (IH)

Table 8.1 Berry phases α in units of π obtained without (top panel) and with (bottom panel) optimization for the exchange of two quasiholes in the Kapit-Mueller (KM) model and in the interacting Hofstadter (IH) model with open boundary conditions. The values of α obtained from the two methods that are closest to the ideal value $1/2$ are indicated with bold text.

discretize this process into N_{steps} steps. As described before, we also include auxiliary potentials on all the other sites, and we optimize these for each value of λ. By repeating this procedure, we make a complete exchange of the two anyons.

We compute the Berry phase factor $e^{i\pi\alpha}$ for the exchange both with and without optimization. For the case without optimization, we start out with the trapping potentials at the lattice sites (2,2) and (5,5). For the case with optimization, we start from the situation depicted in Fig. 8.2(d). Here, we assume the adiabatic limit and follow the method in Ref. [40] to compute the Berry phase factor. Specifically, we compute the ground state at each step by exact diagonalization, and we fix the phase of the ground state by requiring that the overlap between the wavefunction at the current step and the wavefunction at the previous step is real. The Berry phase can then be read off by comparing the phase of the final state to the phase of the initial state. The expected value for the case of well separated anyons is $\alpha = 1/2$ and hence $e^{i\pi\alpha} = i$.

In Table 8.1, we compare the cases with and without optimization for both the Kapit-Mueller model and the interacting Hofstadter model. We find that a small value of N_{steps} is enough to reach the regime, in which the Berry phase α is not sensitive to the precise choice of N_{steps}. For the case without optimization, we find that a high strength of the potential is required ($V_0 \geq 100$) to obtain reasonably good results. For the case with optimization, however, a quite weak potential $V_0 < 10$ is sufficient. In addition, the Berry phases obtained with optimization are much closer to the ideal value than the ones obtained without optimization. This can also be seen in Fig. 8.3(a), which shows the effect of the strength V_0 on the numerical value of $e^{i\pi\alpha}$ for the case with optimization. In general, the larger V_0 is, the better the $e^{i\pi\alpha}$ is. However, the numerical result saturates when V_0 exceeds a threshold value and thus strong V_0 is not necessary.

8.2.4 Robustness

In an experiment, the desired potentials on different lattice sites may not be exactly achieved, and we therefore test how robust the scheme is against possible errors in the auxiliary potentials as shown in Fig. 8.3(b). We consider potentials $V_i' = V_i^{\mathrm{opt}}(1+\delta)$ that are slightly perturbed away from the optimized potentials V_i^{opt} due to an error δ coming from uncertainties in the operations in the experiment. First, we show the robustness of our method with respect to screening of the anyons by plotting the fitness F as a function of the size of the error δ introduced. We observe that when the error is reasonably small, say $\delta < 10^{-2}$, the fitness function is quite small $F < 10^{-4}$, thus the anyons are still very well screened. Secondly, we demonstrate the robustness with respect to the relative error of the Berry phase once δ is introduced. The relative error of the Berry phase is defined as

$$\Delta = \frac{\left| \mathrm{Im}(e^{i\pi\alpha_\delta}) - \mathrm{Im}(e^{i\pi\alpha}) \right|}{\mathrm{Im}(e^{i\pi\alpha})}, \tag{8.9}$$

where α_δ is the Berry phase obtained with the error. Notice that the relative error is small $\Delta \sim O(10^{-2})$ when the error is reasonably small $\delta \leq 10^{-2}$.

8.2.5 Braiding within a finite time

The computation above to determine the Berry phase assumes that the anyons are exchanged infinitely slowly to ensure that the ground state returns to itself with no mixing with the excited states. It is, however, also relevant to know how slowly the anyons should be moved to be close to adiabaticity, and we therefore now consider the outcome when the braiding is completed within a fixed time T.

Ideally one would start with an initial wavefunction $|\psi(0)\rangle$ and achieve the wavefunction at time t through the expression

$$|\psi(t)\rangle = U(t)|\psi(0)\rangle, \tag{8.10}$$

where $U(t) = \mathscr{T}\exp[-i\int_0^t H(t')dt']$ is the time evolution operator, $\mathscr{T}$ stands for time ordering, and $H(t) = H_0 + H_p(t)$ is the time-dependent Hamiltonian. Numerically, however, one

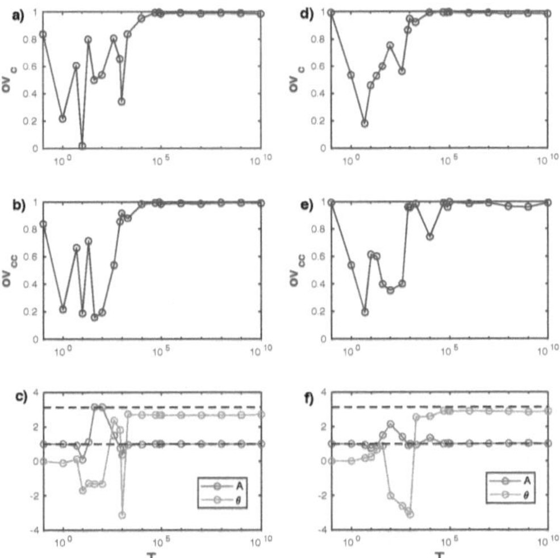

Fig. 8.4 Braiding results for the exchange of two anyons in the Kapit-Mueller model. The
left (right) hand side of the figure is without (with) optimization. (a,d) The overlap $\mathrm{ov_c} =$
$|\langle\psi(0)|\psi(T)\rangle_{\mathrm{clockwise}}|$ between the initial and final states when the anyons are exchanged
in the clockwise direction as a function of the duration T of the exchange. (b,e) The same
quantity, but for a counterclockwise exchange. (c,f) Amplitude A and phase θ of the ratio R
of the overall phases computed for the clockwise and counterclockwise cases. The dashed
lines show the ideal values $A = 1$ and $\theta = \pi$.

must split the unitary operator into a finite number of steps $\mathcal{N}$, and hence we use the relation

$$U(t) = \lim_{\mathcal{N} \to \infty} \prod_{j=0}^{\mathcal{N}-1} e^{-iH(j\Delta t)\Delta t}. \tag{8.11}$$

In the above expression, $\Delta t = t/\mathcal{N}$ and we consider large $\mathcal{N}$ to ensure convergence.

For the case without optimization, we move the potential of strength V from say site a to site b in $\mathcal{N}$ steps using

$$H_p(j\Delta t) = \left[\cos^2 \left(\frac{j\pi}{2\mathcal{N}} \right) n_a + \sin^2 \left(\frac{j\pi}{2\mathcal{N}} \right) n_b \right] V \tag{8.12}$$

where $j \in \{0, 1, \ldots, \mathcal{N} - 1\}$. We compute this for the Kapit-Mueller model and use the same braiding path as shown in Fig. 8.2(a).

For the case with optimization, we reuse the optimized potentials computed for the Kapit-Mueller model with $N_{\text{steps}} = 4$. Here, we need additional steps to ensure convergence, and we hence use $\mathcal{N} N_{\text{steps}}$ steps to move the anyons one lattice spacing. Instead of doing an optimization for each of these smaller steps, we smoothly interpolate between the potentials already computed for two successive steps i and $i+1$ through

$$H_p(i\mathcal{N}\Delta t + j\Delta t) = \cos^2 \left(\frac{j\pi}{2\mathcal{N}} \right) H_p^i + \sin^2 \left(\frac{j\pi}{2\mathcal{N}} \right) H_p^{i+1}, \tag{8.13}$$

where $i \in \{0, 1, \ldots, N_{\text{steps}} - 1\}$, $j \in \{0, 1, \ldots, \mathcal{N} - 1\}$ and H_p^i are the optimized potentials computed for the ith step. By this procedure, we achieve a total of $4\mathcal{N}$ steps quite easily which would otherwise require a huge numerical effort if we had to run the optimization algorithm for each of the $4\mathcal{N}$ steps.

We repeat this procedure until the anyons have been exchanged, and we let T denote the total time used for the exchange. When T is sufficiently large, the process is adiabatic, and the phase acquired during the exchange is $\langle \psi(0)|\psi(T)\rangle$. This phase, however, has contributions from both the dynamical phase and the statistical phase due to the exchange of the anyons. The dynamical phase is prone to numerical error accumulated due to variations in the instantaneous energies at each point along the braiding path. We also note that exchanging the two anyons clockwise and counterclockwise results in the same dynamical phase and hence also the same numerical error. We can hence get rid of the dynamical phase by computing the ratio

$$R = \frac{\langle \psi(0)|\psi(T)\rangle_{\text{clockwise}}}{\langle \psi(0)|\psi(T)\rangle_{\text{counterclockwise}}}. \tag{8.14}$$

In general, $R = A \exp(i\theta)$ and for the ideal case with well-separated anyons, $A = 1$ and $\theta = \pi$.

We compute θ, A, and the overlap between the initial and final wavefunctions for the cases with and without optimization for different total times T for the Kapit-Mueller model as shown in Fig. 8.4. The time required to achieve adiabaticity for the case with optimization (T_{opt}) is slightly larger compared to the case without optimization (T_{normal}), and roughly $T_{\mathrm{opt}} \approx 5 T_{\mathrm{normal}}$. This is possibly due to the smaller energy gap to the first excited state for the case with optimization compared to the case without optimization. We see that the braiding phase is better for the case with optimization due to the better screening of the anyons and we believe a strict optimization at each step will result in a more accurate braiding phase.

8.3   Conclusion

In this chapter, we first demonstrated in a toy model that squeezing anyons and braiding them is possible under some assumptions. We then considered realistic FQH lattice Hamiltonians and incorporated global optimization scheme to trap and squeeze Abelian anyons on a square lattice with open boundary conditions. Specifically, we tuned the auxiliary potentials on different lattice sites through the algorithm to ensure strict localization of anyons. Next, we braided these squeezed anyons by crawling them like snakes with minimal overlap between them. The braiding statistics computed numerically for the adiabatic case through optimization was much better than the case without any optimization. We find that the squeezed anyons need to move at a slower speed to be close to the adiabatic limit (by a factor of about five for the considered example). We note, however, that to get better results for the Berry phase for the normal anyons, one would need to separate the anyons further, which would also increase the time needed for the braiding. An important property of anyon braiding is the topological robustness against small, local disturbances. In the scheme used here, we only optimize the shape of the anyons and not the Berry phase itself, and we rely on adiabatic time evolution to obtain the Berry phase. In this way the topological robustness is maintained. Using a relatively simple physical quantity for the optimization can also be an advantage for experiments. For instance, by measuring just the density, one can find the optimal potentials to squeeze the anyons even if one does not know the model for the considered system exactly.

Chapter 9

Detection of topological quantum phase transitions through quasiparticles

In conventional quantum phase transitions, a local order parameter distinguishes between different phases leading to the success of Landau's theory. However, as discussed in Chapter 2, there exist topologically ordered phases that do not fit under the Landau paradigm and are usually characterized by the presence of long range entanglement. This implies that the quantum state cannot be deformed into a product state through any local unitary evolution [131]. This long-range entanglement can lead to unique properties such as the ground state degeneracy on a particular manifold and fractionalized excitations referred to as anyons. Due to the lack of local order parameters it is usually a hard task to uniquely identify a topologically ordered phase.

However, different probes are available to detect topological phase transitions, such as ground state degeneracy [132], many-body Chern number [133–135], spectral flow [136–138], entanglement spectrum [138–141], topological entanglement entropy [142–144], and fidelity [145]. The first three assume particular boundary conditions. The entanglement based probes have been tested for regular structures in two dimensions, but it is not clear how and whether they can be applied in highly irregular systems. Fidelity cannot be used if the Hilbert space itself changes as a function of the parameter. There are hence systems that cannot be handled currently. In addition, it is desirable to find probes that are less costly numerically. There is hence a strong demand for identifying further probes.

In this Chapter, we exploit the fact that anyons are the correct quasiparticles only in the topologically ordered phase, and hence use them as detectors of topological quantum phase transitions (TQPT). Our starting point is to modify the Hamiltonian locally to generate quasiparticles at well-defined positions in the ground state. In the simplest case, this can be done by adding a potential, while in more complicated systems, it may require some ingenuity [35]. We then study the properties of the quasiparticles as a function of the parameter. When the two phases do not support the same set of quasiparticles, a change is seen at the phase transition. The method can be applied for all types of anyons, as long as there is an appropriate way to create the anyons, and it does not require a particular choice of boundary conditions or a particular lattice structure. The method therefore also applies to, e.g., disordered systems, fractals, and quasicrystals.

We test the method on concrete examples, namely phase transitions occurring in a lattice Moore-Read model on a square lattice and on a fractal lattice, in an interacting Hofstadter model in the presence and in the absence of disorder, and in Kitaev's toric code in a magnetic field. Among these models, we include cases, for which the phase transition point is already known, since this allows us to compare with other methods and check the reliability of the anyon approach. For all these examples, we find that it is sufficient to compute a relatively simple property, such as the charge of the anyons, to determine the phase transition point. The computations can therefore be done at low numerical costs. For the Moore-Read model on a square lattice, e.g., a large speed up is found compared to previous computations of the topological entanglement entropy, and this enables us to determine the transition point much more accurately. For the interacting Hofstadter model, we only need two exact diagonalizations for each data point, which is much less than what is needed to compute the many-body Chern number. Finally, for the model on the fractal, we do not know of other methods that could be used for detecting the phase transition.

This chapter is based on the following reference [146].

- " *Quasiparticles as Detector of Topological Quantum Phase Transitions* ", S. Manna, **N. S. Srivatsa**, J. Wildeboer & A. E. B. Nielsen, **Phys. Rev. Research 2, 043443 (2020)**

9.1   Lattice Moore-Read model

The Moore-Read state is a trial wavefunction to describe the plateau at filling factor $5/2$ in the fractional quantum Hall effect [147], and it supports non-Abelian Ising anyons [148]. In this section, we investigate phase transitions that happen in lattice versions of the Moore-Read state on two different lattices as a function of the lattice filling.

9.1.1   Moore-Read model on a square lattice

We investigate a model with a particular type of lattice Moore-Read ground state, which was shown in [149], based on computations of the topological entanglement entropy γ, to

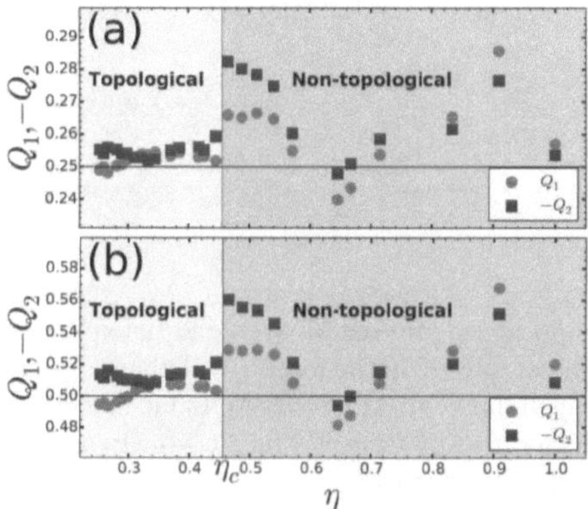

Fig. 9.1 (a) Excess charges Q_1 and Q_2 for the Moore-Read state $|\Psi_a\rangle$ on a square lattice as a function of the flux per site η. In the topological phase, Q_1 and $-Q_2$ are close to the charge of the positive anyon (horizontal line at $1/4$). In the nontopological phase, Q_1 and Q_2 may take any value. The jump away from $1/4$ predicts the transition point $\eta_c \in [0.44, 0.46]$. (b) To test the robustness of the method, we observe that $|\Psi_h\rangle$ gives the same transition point. The Monte Carlo errors are of order 10^{-4}.

exhibit a phase transition as a function of the lattice filling with the transition point in the interval $[1/8, 1/2]$. A more precise value was not determined because γ is expensive to compute numerically, since it involves computing several entanglement entropies, and these are obtained using the replica trick, which means that one works with a system size that is twice as big as the physical system. In fact, for many systems, it is only possible to compute γ for a range of system sizes that are too small to allow for an extrapolation to the thermodynamic limit. Here, we show that the transition point can be found by computing the charge of the anyons. This quantity can be expressed as a classical mean value and is much less expensive to compute. As a result, we can determine the transition point more accurately.

We consider a square lattice with a roughly circular boundary to mimic a quantum Hall droplet. The N lattice sites are at the positions $z_1, \ldots, z_N$, and the local basis on site j is $|n_j\rangle$, where $n_j \in \{0, 1, 2\}$ is the number of bosons on the site. The lattice Moore-Read state $|\Psi_0\rangle$ is defined as [149]

$$|\Psi_x\rangle \propto \sum_{n_1, \ldots, n_N} \Psi_x(n_1, \ldots, n_N) \, |n_1, \ldots, n_N\rangle, \tag{9.1}$$

$$\Psi_x(n_1, \ldots, n_N) = \mathscr{G}_n^x \delta_n \prod_{i<j} (z_i - z_j)^{2n_i n_j} \prod_{i \neq j} (z_i - z_j)^{-\eta n_i},$$

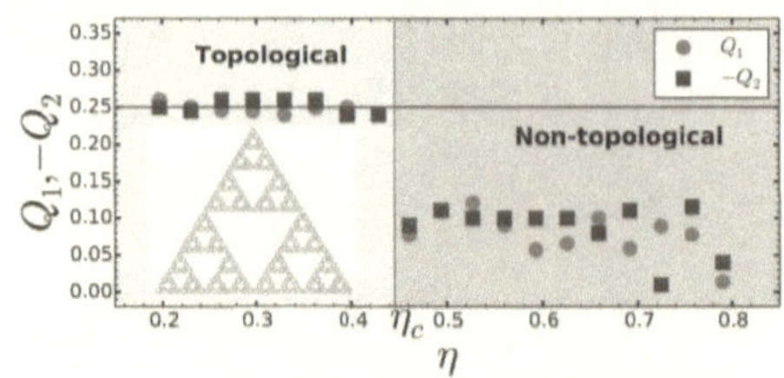

Fig. 9.2 Excess charges Q_1 and Q_2 for the Moore-Read state $|\Psi_a\rangle$ on a fractal lattice (inset) as a function of the flux per site η. In the topological phase, Q_1 and $-Q_2$ are close to the charge of the positive anyon (horizontal line at $1/4$). In the nontopological phase, Q_1 and Q_2 may take any value. The jump away from $1/4$ predicts the transition point $\eta_c \in [0.43, 0.46]$. The Monte Carlo errors are of order 10^{-4}.

where $\mathscr{G}_n^0 = \mathrm{Pf}[1/(z_i' - z_j')]$, $\mathrm{Pf}(\dots)$ is the Pfaffian, z_i' are the positions of the $\mathscr{M}$ singly occupied lattice sites, δ_n is a delta function that enforces the number of particles to be $M \equiv \sum_i n_i = \eta N/2$, and η is the magnetic flux per site. Note that we can vary the lattice filling factor $M/N = \eta/2$ by changing η.

We also introduce the states $|\Psi_a\rangle$ and $|\Psi_b\rangle$ with

$$\mathscr{G}_n^a = 2^{-\frac{\mathscr{M}}{2}} \prod_j (w_2 - z_j)^{-n_j}$$

$$\times \mathrm{Pf}\left[\frac{(z_i' - w_1)(z_j' - w_2) + (z_j' - w_1)(z_i' - w_2)}{(z_i' - z_j')} \right], \tag{9.2}$$

$$\mathscr{G}_n^b = \mathrm{Pf}\left(\frac{1}{z_i' - z_j'} \right) \prod_j (w_1 - z_j)^{n_j} \prod_j (w_2 - z_j)^{-n_j}. \tag{9.3}$$

When the system is in the topological phase, the state $|\Psi_a\rangle$ ($|\Psi_b\rangle$) has an anyon of charge $+1/4$ ($+1/2$) at w_1 and an anyon of charge $-1/4$ ($-1/2$) at w_2 [150]. Few-body parent Hamiltonians for the states can be derived for a range of η values and arbitrary choices of the lattice site positions z_i [150, 149].

When anyons are present in the system, they modify the particle density in local regions around each w_k. Let us consider a circular region with radius R. If R is large enough to enclose the anyon, but small enough to not enclose other anyons, the number of particles missing within the region, which is given by the excess charge

$$\mathscr{Q}_k = -\sum_{i=1}^N \rho(z_i)\, \theta(R - |z_i - w_k|), \tag{9.4}$$

equals the charge of the anyon at w_k. Here, $\theta(\dots)$ is the Heaviside step function and

$$\rho(z_i) = \langle \Psi_x | n_i | \Psi_x \rangle - \langle \Psi_0 | n_i | \Psi_0 \rangle, \quad x \in \{a,b\} \tag{9.5}$$

is the density profile of the anyons. In the nontopological phase, a more complicated density pattern can arise. The expectation is hence that $\mathscr{Q}_k$ is close to the anyon charge in the topological phase and varies with η in the nontopological phase, and we use this to detect the transition.

In Fig. 9.1, we choose $R = |w_1 - w_2|/2$ and $M = 40$, and we vary the number of lattice sites from $N = 316$ to $N = 80$ to achieve different η values in the range $[1/4, 1]$. We observe that the excess charges for $|\Psi_a\rangle$ are $Q_1 \approx -Q_2 \approx 1/4$ for $\eta < \eta_c$ and fluctuate for $\eta > \eta_c$, where $\eta_c \in [0.44, 0.46]$. As a test of the robustness of the approach, we observe that the same transition point is predicted using $|\Psi_b\rangle$. We have also checked that the fact that there is a jump in the excess charges from $\eta \simeq 0.44$ to $\eta \simeq 0.46$ is insensitive to the precise choice of the distance $|w_1 - w_2|$.

9.1.2 Moore-Read model on a fractal lattice

We next consider the Moore-Read model on a fractal lattice. The fractal lattice is not periodic, and we can therefore not apply methods, such as ground state degeneracy, spectral flow, or many-body Chern number computations to detect a possible phase transition. Methods based on entanglement computations do also not apply, since we do not have a thorough understanding of entanglement properties of topological many-body states on fractal lattices. Fidelity cannot be used either, since the Hilbert space changes when the considered parameter changes. Quasiparticle properties, on the contrary, can detect a transition, as we will now show.

Lattice Laughlin fractional quantum Hall models were recently constructed on fractals [151], and we here consider a similar construction for the Moore-Read state. Specifically, we define the state $|\Psi_a\rangle$ on a lattice constructed from the Sierpiński gasket with $N = 243$ triangles by placing one lattice site on the center of each triangle. In Fig. 9.2, we vary the particle number $M \in [24, 96]$ to have different $\eta \in [0.19, 0.79]$ values and plot the excess charges as a function of η. The excess charges are $Q_1 \approx -Q_2 \approx 1/4$ for $\eta < \eta_c$ and fluctuate for $\eta > \eta_c$, which reveals a phase transition at the transition point $\eta_c \in [0.43, 0.46]$.

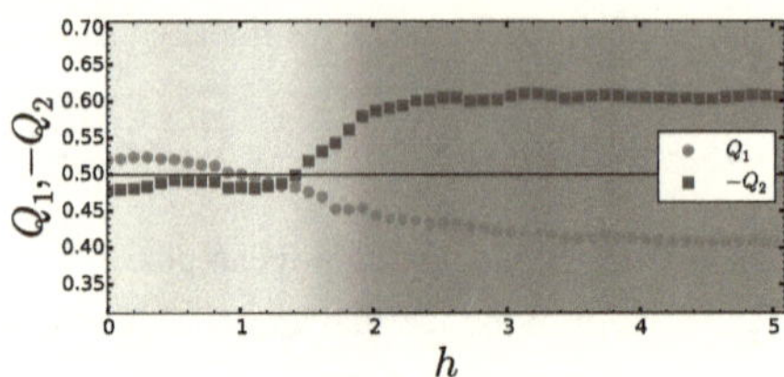

Fig. 9.3 Excess charges Q_1 and Q_2 as a function of the disorder strength h for the interacting Hofstadter model with $M = 3$, $N = 24$, and $\alpha = 0.25$. In the topological phase, $Q_1 \approx -Q_2 \approx 1/2$ (horizontal line), and the observed change away from this value predicts the transition point $h_c \simeq 1.5$. We average over 2000 statistically independent disorder realizations for each h to ensure convergence of the data.

9.2 Disordered interacting Hofstadter model

As another example, we study a Hofstadter model for hardcore bosons on a square lattice. The clean model is known to host a topological phase for low enough lattice filling factor [152, 153], and here we investigate the effect of adding a disordered potential. The system sizes that can be reached with exact diagonalization are too small to allow for a computation of the topological entanglement entropy. Instead, we use the anyon charges to show that the system undergoes a phase transition as a function of the disorder strength. This gives a large speed up in computation time compared to the many-body Chern number computations that were done in [152]. This is so, because it only takes two exact diagonalizations per date point to get the anyon charges, while the Chern number computation involves a large number of exact diagonalizations per data point, corresponding to a grid of twist angles in two dimensions.

The Hofstadter model describes particles hopping on a two-dimensional square lattice in the presence of a magnetic field perpendicular to the plane. Hopping is allowed between nearest neighbor sites, and the magnetic field is taken into account by making the hopping amplitudes complex. Whenever a particle hops around a closed loop, the wavefunction acquires a phase, which is equal to the Aharonov-Bohm phase for a charged particle encircling the same amount of magnetic flux.

We take open boundary conditions and add interactions by considering hardcore bosons. For a lattice with $N = L_x \times L_y$ sites, the Hamiltonian takes the form

$$H_0 = -\sum_{x=1}^{L_x-1} \sum_{y=1}^{L_y} \left(c_{x+1,y}^\dagger c_{x,y} e^{-i\pi\alpha y} + \text{H.c.} \right) \tag{9.6}$$

$$-\sum_{x=1}^{L_x} \sum_{y=1}^{L_y-1} \left(c_{x,y+1}^\dagger c_{x,y} e^{i\pi\alpha x} + \text{H.c.} \right) + \sum_{x=1}^{L_x} \sum_{y=1}^{L_y} h_{x,y} n_{x,y},$$

where $c_{x,y}$ is the hardcore boson annihilation operator and $n_{x,y} = c_{x,y}^\dagger c_{x,y}$ is the number operator acting on the lattice site at the position (x,y) with $x \in \{1,\ldots,L_x\}$ and $y \in \{1,\ldots,L_y\}$. If a particle hops around a plaquette, the phase acquired is $2\pi\alpha$, so α is the flux through the plaquette. We here consider the case, where the number of flux units per particle is two, i.e. $N\alpha/M = 2$. The last term in (9.6) is the disordered potential, and $h_{x,y} \in [-h,h]$ is drawn from a uniform distribution of width $2h$, where h is the disorder strength.

In the clean model, it is well-known that one can trap anyons in the ground state by adding a local potential with a strength that is sufficiently large compared to the hopping amplitude [154? , 155]. We here choose

$$H_V = V n_{x_1,y_1} - V n_{x_2,y_2}, \quad (x_1,y_1) \neq (x_2,y_2), \tag{9.7}$$

where $V \gg 1$. This potential traps one positively (negatively) charged anyon at the site (x_1,y_1) $((x_2,y_2))$.

We use the excess charge in a region around the sites (x_1,y_1) and (x_2,y_2) to detect the phase transition. We define the density profile as

$$\rho(x+iy) = \langle n_{x,y} \rangle_{H_0+H_V} - \langle n_{x,y} \rangle_{H_0}, \tag{9.8}$$

where $\langle n_{x,y} \rangle_{H_0+H_V}$ is the particle density, when the trapping potential is present, and $\langle n_{x,y} \rangle_{H_0}$ is the particle density, when the trapping potential is absent. The excess charge is then defined as in (9.4) with $w_1 = x_1 + iy_1$ and $w_2 = x_2 + iy_2$. Here, we choose R such that the circular region includes all sites up to the second nearest neighbor sites. The absolute value of the excess charge should be close to $1/2$ in the topological region, while it can take any value and may vary with h in the nontopological region.

We choose a point, which is deep in the topological phase for $h = 0$, namely $M = 3$, $N = 24$, and $\alpha = 0.25$, and plot the excess charges as a function of the disorder strength h in Fig. 9.3. We observe that Q_1 and $-Q_2$ are close to $1/2$ up to $h \simeq 1.5$, while the excess

M	N	$L_x \times L_y$	α	$\dim(\mathcal{H})$	Q_+	Q_-
2	24	6×4	0.167	276	0.491	0.507
3	28	7×4	0.214	3276	0.476	0.522
3	24	6×4	0.250	2024	0.519	0.478
4	28	7×4	0.286	20475	0.521	0.475
4	24	6×4	0.333	10626	0.475	0.520
5	28	7×4	0.357	98280	0.465	0.532
6	32	8×4	0.375	906192	0.466	0.533
7	36	6×6	0.389	8347680	0.380	0.610
5	25	5×5	0.400	53130	0.332	0.666
5	24	6×4	0.417	42504	0.348	0.650
6	28	7×4	0.429	376740	0.250	0.747
6	25	5×5	0.480	177100	0.278	0.719
7	28	7×4	0.500	1184040	0.284	0.714
7	25	5×5	0.560	480700	0.402	0.595
7	24	6×4	0.583	346104	0.440	0.557

Table 9.1 We show here the different choices we make for the number of particles M, the shapes and sizes $N = L_x \times L_y$ of the lattices, and the fluxes per plaquette $\alpha = 2M/N$. The quantity $\dim(\mathcal{H})$ is the dimension of the corresponding Hilbert spaces. We display the data for the absolute values of the excess charges Q_+ and Q_-. There is a significant change in Q_+ and Q_-, when going from $\alpha = 0.375$ to $\alpha \simeq 0.389$.

charges deviate more from $1/2$ for $h > 1.5$. The data hence predict the phase transition to happen at $h_c \simeq 1.5$.

We can also put the disorder strength to $h = 0$ and study the clean model as a function of the magnetic flux α. Specifically, we vary α and the lattice filling M/N, while keeping the flux per particle $N\alpha/M$ fixed. For this case, it was found in [152] that there is a phase transition at $\alpha_c \in [0.375, 0.400]$. We take values of M and N (see Tab. 9.1), which are numerically accessible for exact diagonalization, and for each choice $\alpha = 2M/N$. Figure 9.4 shows that Q_1 and $-Q_2$ are quite close to $1/2$ for α values up to 0.375, but for higher α they deviate much more from $1/2$. The data hence predict the transition point $\alpha_c \in [0.375, 0.389]$, which is consistent with the result in [152].

9.3 Toric code in a magnetic field

To test the applicability of the method outside the family of chiral fractional quantum Hall models, we next study Kitaev's toric code [156, 157] on a square lattice with periodic boundary conditions. This system exhibits a $\mathbb{Z}_2$ topologically ordered phase, and it is

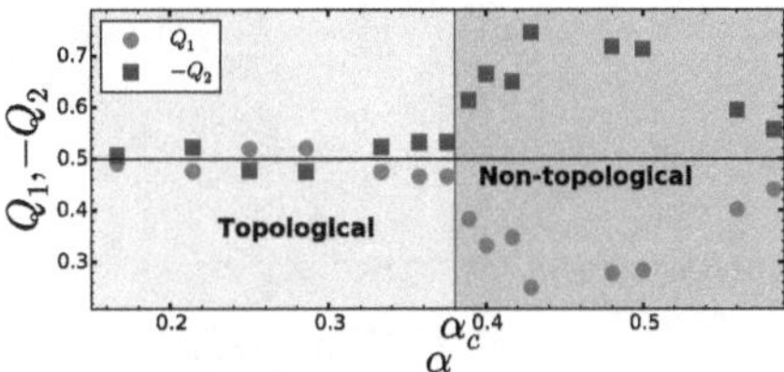

Fig. 9.4 Excess charges Q_1 and Q_2 as a function of the magnetic flux per plaquette α for the interacting Hofstadter model without disorder. In the topological phase, $Q_1 \approx -Q_2 \approx 1/2$, and the observed change away from this value predicts the transition point $\alpha_c \in [0.375, 0.389]$.

known that a sufficiently strong, uniform magnetic field drives the system into a polarized phase [158–162]. Here we show that anyons inserted into the system are able to detect this phase transition, and the obtained transition points agree with earlier results based on perturbative, analytical calculations and tensor network studies. Our computations rely on exact diagonalization for a system with 18 spins and are hence quite fast to do numerically. We find that the anyons are significantly better at predicting the phase transition point than the energy gap closing for the same system size.

The toric code has a spin-$1/2$ on each of the edges of the $N_x \times N_y$ square lattice. The Hamiltonian

$$H_{\mathrm{TC}} = -\sum_p B_p - \sum_v A_v, \quad B_p = \prod_{i \in p} \sigma_i^z, \quad A_v = \prod_{i \in v} \sigma_i^x, \tag{9.9}$$

is expressed in terms of the Pauli operators σ_i^x and σ_i^z, which act on the $N = 2N_x N_y$ spins. The sums are over all plaquettes p and vertices v of the lattice. B_p acts on the spins on the four edges surrounding the plaquette p, and A_v acts on the spins on the four edges connecting to the vertex v.

H_{TC} is exactly solvable, and the four degenerate ground states are eigenstates of B_p and A_v with eigenvalue 1. States containing anyons are obtained by applying certain string operators to the ground states. The string operator either changes the eigenvalue of two A_v operators to -1 or the eigenvalue of two B_p operators to -1. In the former case, two electric excitations e_v are created, and in the latter case two magnetic excitations m_p are created. The wavefunction acquires a minus sign if one m_p is moved around one e_v, and the excitations are hence Abelian anyons.

Here, we instead modify the Hamiltonian, such that anyons are present in the ground states. The ground states of the Hamiltonian $H_m \equiv H_{\mathrm{TC}} + 2B_{p_1} + 2B_{p_2}$ have one m_p on each of the plaquettes p_1 and p_2. Similarly, the ground states of the Hamiltonian $H_e \equiv H_{\mathrm{TC}} + 2A_{v_1} + 2A_{v_2}$ have one e_v on each of the vertices v_1 and v_2. We drive the system through a phase transition

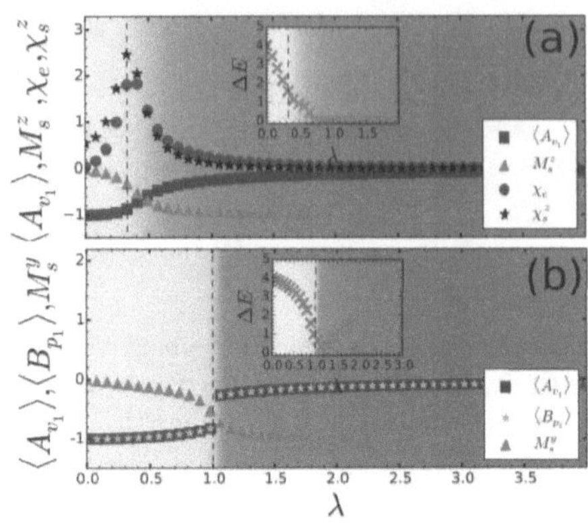

Fig. 9.5 (a) The toric code with a magnetic field of strength λ in the z direction undergoes a phase transition at $\lambda_c \simeq 0.33$. The transition is seen in $\langle A_{v_1} \rangle$, which detects the anyons in the ground states of $H_e + H_\lambda^z$, and in $M_s^z = \frac{1}{N} \langle \sum_i \sigma_i^k \rangle$, which is the magnetization per spin for the ground states of $H_{\mathrm{TC}} + H_\lambda^z$. We also show $\chi_e = \partial \langle A_{v_1} \rangle / \partial \lambda$ and $\chi_s^z = -\partial M_s^z / \partial \lambda$. (b) When the magnetic field is in the y direction, the phase transition happens at $\lambda_c = 1$. We plot $\langle A_{v_1} \rangle$ for $H_e + H_\lambda^y$ and $\langle B_{p_1} \rangle$ for $H_m + H_\lambda^y$ as well as the magnetization per spin $M_s^y = \frac{1}{N} \langle \sum_i \sigma_i^y \rangle$ for $H_{\mathrm{TC}} + H_\lambda^y$. The closing of the energy gap ΔE between the ground state (fourfold degenerate) and the first excited state of $H_{\mathrm{TC}} + H_\lambda^k$ shown in the insets does not accurately predict the transition points. We use a 3×3 lattice with 18 spins in all cases, and the color gradient from yellow (topological phase) to green (nontopological phase) is plotted according to the value of $\langle A_{v_1} \rangle$.

by adding a magnetic field $H_\lambda^k = \lambda \sum_i \sigma_i^k$ in the k-direction with strength λ. When λ is large enough, it is energetically favorable to polarize all the spins, and the system is no longer topological.

Previous investigations, based on perturbative, analytical calculations and tensor network studies [158–160], have shown that $H_{\mathrm{TC}} + H_\lambda^z$ has a second order phase transition at $\lambda_c \simeq 0.33$, while $H_{\mathrm{TC}} + H_\lambda^y$ has a first order phase transition at $\lambda_c = 1$. The magnetization per spin computed using exact diagonalization (Fig. 9.5) gives similar values for the transition points.

We now use anyons to detect the transition. We study $\langle A_{v_1} \rangle = \langle A_{v_2} \rangle$ for the Hamiltonian $H_e + H_\lambda^k$ and $\langle B_{p_1} \rangle = \langle B_{p_2} \rangle$ for the Hamiltonian $H_m + H_\lambda^k$. $\langle A_{v_1} \rangle = -1$ and $\langle B_{p_1} \rangle = -1$ signify the presence of the anyons. In the fully polarized phase, both $\langle A_{v_1} \rangle$ and $\langle B_{p_1} \rangle$ vanish when $k = y$ and $\langle A_{v_1} \rangle$ vanishes and $\langle B_{p_1} \rangle \to +1$ when $k = z$.

The transition seen in $\langle A_{v_1} \rangle$ for the ground states of $H_e + H_\lambda^z$ in Fig. 9.5(a) is consistent with $\lambda_c \simeq 0.33$. Both the transition point and the width of the transition region, which is due to finite size effects, are comparable to the same quantities obtained from the magnetic order parameter. The anyons predict the transition point more accurately than the energy gap closing, which, for the same system size, happens only at $\lambda \simeq 0.7$.

$\langle B_{p_1} \rangle$ is not suitable for detecting the transition because B_{p_1} commutes with all terms in $H_m + H_\lambda^z$. All energy eigenstates are therefore also eigenstates of B_{p_1} with eigenvalue ± 1. As a result, $\langle B_{p_1} \rangle$ only measures whether the ground states have B_{p_1} eigenvalue $+1$ or -1. The first transition to a ground state with B_{p_1} eigenvalue $+1$ happens around $\lambda \simeq 2.08$, but this does not exclude gap closings at smaller λ values. These problems do not occur for $\langle A_{v_1} \rangle$, since A_{v_1} does not commute with $H_e + H_\lambda^z$.

The transition seen in $\langle A_{v_1} \rangle$ ($\langle B_{p_1} \rangle$) in Fig. 9.5(b) for the ground states of $H_e + H_\lambda^y$ ($H_m + H_\lambda^y$) is consistent with $\lambda_c = 1$, and the transition is sharper than for the magnetic order parameter. The anyons also better predict the transition point than the energy gap closing, which happens around $\lambda \simeq 1.2$ for the same system size.

9.4 Conclusion

In this chapter, we have shown that properties of quasiparticles are an interesting tool to detect topological quantum phase transitions. The approach is to trap anyons in the ground state and study how their properties change, when the system crosses a phase transition. For several quite different examples, we have demonstrated, however, that the phase transitions can be detected by observing a simple property, such as the charge of the anyons. This means that the phase transition points can be computed at low numerical costs.

The approach suggested here to detect topological quantum phase transitions is particularly direct, since, to fully exploit the interesting properties of topologically ordered systems, one needs to be able to create anyons in the systems and detect their properties. In the interacting Hofstadter model, the anyons can be created by adding a local potential, and the charge of the anyons used to detect the phase transition can be measured by measuring the expectation value of the number of particles on each site. Both of these can be done in experiments with ultracold atoms in optical lattices [163?]. For the toric code model, the complexity of generating the Hamiltonians including the terms creating the anyons is about the same as generating the Hamiltonian without these terms, and first steps towards realizing the Hamiltonian in experiments have been taken [164–166].

The ideas presented in this chapter can be applied as long as the two phases support quasiparticles with different properties and one can find suitable ways to create the quasiparticles.

Chapter 10

Conclusions and outlooks

Interacting quantum many-particle systems in lower dimensions, display interesting emergent phases as a result of strong quantum correlations. These phases are tied to the many-body eigenstates of the model under investigation. For example, topological phases, which are characterized by long range entanglement and anyonic excitations, are traditionally known to occur in gapped quantum ground states which have a particular degeneracy on a given manifold [1].

The study of many-body quantum systems has hence largely focussed on the ground-state properties and low-energy excitations, with an implicit assumption that highly excited states of generic non-integrable models are void of interesting features. This is because, highly excited states of such systems were known to be thermal and conform to the eigenstate thermalization hypothesis (ETH) [8, 9]. Another reason that people have not payed that much attention to excited states is that they are often difficult to prepare. However, through two mechanisms, many-body localization (MBL) [10] and quantum many-body scars [12], it has become clear that eigenstates even at high energy densities may violate ETH and host interesting phases which can also be topological.

This book focusses on aspects of topological phases that occur in the quantum many-body ground states and also explore interesting non-ergodic phases in high energy eigenstates through disorder and quantum scarring. Below, we will briefly motivate and outline the conclusion from various chapters.

Fractional quantum Hall (FQH) effect is a prime example of topological order that showed the limits of Landau's theory [2]. A remarkable feature of FQH is the possibility of fractionalized excitations called anyons with exotic exchange statistics interpolating between fermions and bosons. Lattice is a promising platform for realizing FQH phases and also to create and manipulate anyons in a controlled fashion. One of the intentions of this book is to contribute to the ongoing research in the area of lattice FQH and anyons, which, also opens up interesting research directions as discussed below.

Conformal field theory (CFT) has been a useful tool to construct analytical states with chiral topological order which is expected for FQH states [5]. Also, few-body, non-local parent Hamiltonians have been constructed for these analytical states and these are different from the fractional Chern insulator models. In some cases, with much symmetry, deformations of these models with fine tuned coefficients, leading to local Hamiltonians with ground states bearing expected topological properties have been shown. However, a general procedure to arrive at more tractable local Hamiltonians constructed from CFT is missing. In Chapter 7, we showed a general way of constructing local versions of lattice FQH Hamiltonians constructed from CFT that does not rely on optimization, symmetries, or choice of lattice geometry. In contrast to the earlier methods, we do not truncate the Hamiltonian directly, which does not give an unequivocal result. We rather truncate the operator that annihilates the analytical state of interest and then build the Hamiltonian. Hence, this procedure also applies naturally to periodic boundary conditions. We numerically estimated the radius upto which the operator that annihilates the state needs to be truncated to achieve a local Hamiltonian with similar topological properties. With the aid of exact diagonalization, we tested the success of this approach on lattice bosonic Laughlin models. As a promising future direction, it would be worth to also apply this truncation scheme to lattice FQH models at other filling fractions (such as the Moore Read model) and also to model Hamiltonians with anyons. It would be interesting to test how large the local regions need to be to stabilize the correct topological order in the ground states for these other models derived from CFT. In the latter part of this chapter, we also showed that CFT can be used to obtain local models in 1D with universal quantum critical properties.

Excitations in fractional quantum Hall systems are anyons [33] with fractional statistics that interpolate between fermions and bosons. Anyons may be created in lattice FQH models through trapping potentials. Their presence is evident from the modified particle density in the ground state in local regions around the trapping potential. In lattice models, it has been numerically observed that the anyons are typically spread over a few lattice sites around the position of trapping potentials. Given the small lattice sizes that can be handled through exact diagonalization, a larger braiding path may not be chosen to avoid overlap between the anyons. It becomes an even harder task to braid the anyons on a lattice with open boundaries which is crucial also from an experimental view point. In Chapter 8, we provided a way around the issues mentioned above by instead squeezing the Abelian anyons and braid them by crawling them like snakes. We chose simple lattice FQH models and used an optimization protocol to tune potentials on lattice sites to squeeze Abelian anyons with near zero overlap.

We also showed numerically that the braiding phase achieved by squeezing anyons through optimized potentials is better in comparison to the case where a single trapping potential is used to trap the anyons. Complementarily, we also showed that squeezing anyons and braiding is possible through analytical states derived from CFT under certain assumptions. Applying the similar protocol to trap and squeeze non-Abelian anyons on lattice and braid them is certainly a possible future direction of study.

Since topological order cannot be characterized by local order parameters, a different set of indicators need to be invoked to detect topological quantum phase transitions (TQPT). Although there are quite a few probes to detect a topological phase, there are places where some of these probes fail to apply, become numerically expensive or require special conditions. For instance, it is not clear how the entanglement-based probes may be applied in highly irregular systems, fidelity cannot be used if the Hilbert space itself changes as a function of parameter and, probes such as ground state degeneracy, many-body Chern number and entanglement spectrum require particular boundary conditions. In Chapter 9, we showed that quasiparticles may be used to detect TQPTs quite efficiently in five rather different examples. Since a topologically ordered phase is known to support anyons, in some of the examples we considered, the approach was to trap anyons in the ground state and study how their properties change when the system crosses a phase transition. Specifically, we used the charge of the anyon to detect the transition from a topological to a trivial phase. The advantages of this method are, it applies independent of boundary conditions or a particular lattice choice, is a numerically cheap probe (for example, it typically requires much fewer exact diagonalizations compared to computing the Chern number) and applies well to disordered and irregular systems. As a prospective future research, it would be interesting to explore methods to trap quasi particles in frustrated magnetic models and explore parameter regimes where quantum spin liquid phases have been speculated.

One of the yet other main contributions of this book is in the area of non-equilibrium physics. We showed that, by exploiting symmetry and disorder, one can construct models with novel sorts of non-thermal phases in its eigenstates. We also provided with constructions of two different quantum scarred models with chiral and non-chiral topological order. We briefly describe below the conclusions and possible future directions from rest of the chapters in this book.

Symmetries are known to constrain the physics of MBL [11]. For example, it is known that SU(2) symmetry in a model does not allow the eigenstates to be area law entangled and thus prohibiting a conventional MBL phase [71]. In Chapter 4, we showed that an emergent SU(2) symmetry in the ground state of a disordered model protects it from MBL while not restricting the other states from localizing. Specifically, we showed that the ground state remains quantum critical with Luttinger liquid properties, while the excited states display a transition from an ergodic to a glassy MBL phase with disorder. We also demonstrated how the quantum critical ground state may be lifted to the middle of the spectrum of glassy MBL states, which then would correspond to a situation of weak violation of MBL. One possible interesting direction would be to construct disordered models with multiple states with emergent symmetry being protected from MBL. This would then have the flavour of an inverse situation of the celebrated quantum many-body scars. In Chapter 5, we demonstrated numerically the emergence of a non-ergodic but also non-mbl phase in a disordered SU(2) invariant model with non-local interactions. Exploring non-ergodic phases beyond MBL and studying transport properties in such models is also an interesting research direction that one might pursue.

In Chapter 6, we showed two constructions of scarred models with topological order. In the first case, we exploited the existence of multiple operators constructed from CFT that annihilate the scar states to build Hamiltonians with a free parameter to pin the scar state at any desired position in the many-body spectrum. The scar state was the discretized version of the bosonic Laughlin state with quantum critical properties in 1D and chiral topological order in 2D. These models were defined on slightly disordered lattices to ensure that the spectrum is generic. We showed that in the presence of disorder, the scar state in 2D is topologically ordered by being able to create anyons with the correct charge. Since these models are non-local, it would be useful to invoke the truncation procedure discussed in Chapter 7 to build local versions of the scarred models. Further, it would be interesting to build models with scar states corresponding to anyons and perform braiding even at finite energy densities. As a second construction, we showed a general strategy to construct classes of frustration free models with quantum scars. Specifically we applied this to quantum dimer models on a kagome lattice and showed the existence of an analytically known scar state with non-chiral topological order. It would be worth to study the fate of scarred eigenstates in these models under generic perturbations.

References

[1] X.-G. Wen, "Topological order: From long-range entangled quantum matter to a unified origin of light and electrons," *ISRN Condensed Matter Physics*, vol. 2013, p. 198710, Mar 2013.

[2] R. B. Laughlin, "Anomalous quantum hall effect: An incompressible quantum fluid with fractionally charged excitations," *Phys. Rev. Lett.*, vol. 50, pp. 1395–1398, May 1983.

[3] H. L. Stormer, D. C. Tsui, and A. C. Gossard, "The fractional quantum hall effect," *Rev. Mod. Phys.*, vol. 71, pp. S298–S305, Mar 1999.

[4] A. E. B. Nielsen, J. I. Cirac, and G. Sierra, "Laughlin spin-liquid states on lattices obtained from conformal field theory," *Phys. Rev. Lett.*, vol. 108, p. 257206, Jun 2012.

[5] H.-H. Tu, A. E. B. Nielsen, J. I. Cirac, and G. Sierra, "Lattice laughlin states of bosons and fermions at filling fractions 1/q," *New Journal of Physics*, vol. 16, p. 033025, mar 2014.

[6] A. E. B. Nielsen, "Anyon braiding in semianalytical fractional quantum Hall lattice models," *Phys. Rev. B*, vol. 91, p. 041106, Jan 2015.

[7] A. E. B. Nielsen, G. Sierra, and J. I. Cirac, "Local models of fractional quantum hall states in lattices and physical implementation," *Nature Communications*, vol. 4, p. 2864, Nov 2013.

[8] M. Srednicki, "Chaos and quantum thermalization," *Phys. Rev. E*, vol. 50, pp. 888–901, Aug 1994.

[9] J. M. Deutsch, "Quantum statistical mechanics in a closed system," *Phys. Rev. A*, vol. 43, pp. 2046–2049, Feb 1991.

[10] D. A. Abanin, E. Altman, I. Bloch, and M. Serbyn, "Colloquium: Many-body localization, thermalization, and entanglement," *Rev. Mod. Phys.*, vol. 91, p. 021001, May 2019.

[11] A. C. Potter and R. Vasseur, "Symmetry constraints on many-body localization," *Phys. Rev. B*, vol. 94, p. 224206, Dec 2016.

[12] M. Serbyn, D. A. Abanin, and Z. Papić, "Quantum many-body scars and weak breaking of ergodicity," 2020.

[13] R. Moessner and J. E. Moore, *Topological Phases of Matter*. Cambridge University Press, 2021.

[14] F. D. M. Haldane, "Fractional quantization of the hall effect: A hierarchy of incompressible quantum fluid states," *Phys. Rev. Lett.*, vol. 51, pp. 605–608, Aug 1983.

[15] J. K. Jain, "Composite-fermion approach for the fractional quantum hall effect," *Phys. Rev. Lett.*, vol. 63, pp. 199–202, Jul 1989.

[16] M. Hafezi, A. S. Sørensen, E. Demler, and M. D. Lukin, "Fractional quantum hall effect in optical lattices," *Phys. Rev. A*, vol. 76, p. 023613, Aug 2007.

[17] A. S. Sørensen, E. Demler, and M. D. Lukin, "Fractional quantum hall states of atoms in optical lattices," *Phys. Rev. Lett.*, vol. 94, p. 086803, Mar 2005.

[18] R. N. Palmer and D. Jaksch, "High-field fractional quantum hall effect in optical lattices," *Phys. Rev. Lett.*, vol. 96, p. 180407, May 2006.

[19] R. N. Palmer, A. Klein, and D. Jaksch, "Optical lattice quantum hall effect," *Phys. Rev. A*, vol. 78, p. 013609, Jul 2008.

[20] G. Möller and N. R. Cooper, "Composite fermion theory for bosonic quantum hall states on lattices," *Phys. Rev. Lett.*, vol. 103, p. 105303, Sep 2009.

[21] D. Jaksch and P. Zoller, "Creation of effective magnetic fields in optical lattices: the hofstadter butterfly for cold neutral atoms," *New Journal of Physics*, vol. 5, pp. 56–56, may 2003.

[22] E. J. Mueller, "Artificial electromagnetism for neutral atoms: Escher staircase and laughlin liquids," *Phys. Rev. A*, vol. 70, p. 041603, Oct 2004.

[23] F. Gerbier and J. Dalibard, "Gauge fields for ultracold atoms in optical superlattices," *New Journal of Physics*, vol. 12, p. 033007, mar 2010.

[24] N. R. Cooper, "Optical flux lattices for ultracold atomic gases," *Phys. Rev. Lett.*, vol. 106, p. 175301, Apr 2011.

[25] A. Kitaev, "Fault-tolerant quantum computation by anyons," *Annals of Physics*, vol. 303, no. 1, pp. 2–30, 2003.

[26] D. N. Sheng, Z.-C. Gu, K. Sun, and L. Sheng, "Fractional quantum hall effect in the absence of landau levels," *Nature Communications*, vol. 2, p. 389, Jul 2011.

[27] K. Sun, Z. Gu, H. Katsura, and S. Das Sarma, "Nearly flatbands with nontrivial topology," *Phys. Rev. Lett.*, vol. 106, p. 236803, Jun 2011.

[28] E. Tang, J.-W. Mei, and X.-G. Wen, "High-temperature fractional quantum hall states," *Phys. Rev. Lett.*, vol. 106, p. 236802, Jun 2011.

[29] T. Neupert, L. Santos, C. Chamon, and C. Mudry, "Fractional quantum hall states at zero magnetic field," *Phys. Rev. Lett.*, vol. 106, p. 236804, Jun 2011.

[30] G. Moore and N. Read, "Nonabelions in the fractional quantum hall effect," *Nuclear Physics B*, vol. 360, no. 2, pp. 362–396, 1991.

[31] I. Glasser, J. I. Cirac, G. Sierra, and A. E. B. Nielsen, "Exact parent hamiltonians of bosonic and fermionic moore–read states on lattices and local models," *New Journal of Physics*, vol. 17, p. 082001, aug 2015.

[32] A. E. B. Nielsen, G. Sierra, and J. I. Cirac, "Local models of fractional quantum hall states in lattices and physical implementation," *Nature Communications*, vol. 4, p. 2864, Nov 2013.

[33] F. Wilczek, "Quantum mechanics of fractional-spin particles," *Phys. Rev. Lett.*, vol. 49, pp. 957–959, Oct 1982.

[34] I. Glasser, J. I. Cirac, G. Sierra, and A. E. B. Nielsen, "Lattice effects on laughlin wave functions and parent hamiltonians," *Phys. Rev. B*, vol. 94, p. 245104, Dec 2016.

[35] M. Storni and R. H. Morf, "Localized quasiholes and the majorana fermion in fractional quantum hall state at $v = \frac{5}{2}$ via direct diagonalization," *Phys. Rev. B*, vol. 83, p. 195306, May 2011.

[36] S. Johri, Z. Papić, R. N. Bhatt, and P. Schmitteckert, "Quasiholes of $\frac{1}{3}$ and $\frac{7}{3}$ quantum hall states: Size estimates via exact diagonalization and density-matrix renormalization group," *Phys. Rev. B*, vol. 89, p. 115124, Mar 2014.

[37] F. E. Camino, W. Zhou, and V. J. Goldman, "Aharonov-bohm superperiod in a laughlin quasiparticle interferometer," *Phys. Rev. Lett.*, vol. 95, p. 246802, Dec 2005.

[38] F. E. Camino, W. Zhou, and V. J. Goldman, "Transport in the laughlin quasiparticle interferometer: Evidence for topological protection in an anyonic qubit," *Phys. Rev. B*, vol. 74, p. 115301, Sep 2006.

[39] B. Paredes, P. Fedichev, J. I. Cirac, and P. Zoller, "$\frac{1}{2}$-anyons in small atomic bose-einstein condensates," *Phys. Rev. Lett.*, vol. 87, p. 010402, Jun 2001.

[40] E. Kapit, P. Ginsparg, and E. Mueller, "Non-abelian braiding of lattice bosons," *Phys. Rev. Lett.*, vol. 108, p. 066802, Feb 2012.

[41] M. Račiūnas, F. N. Ünal, E. Anisimovas, and A. Eckardt, "Creating, probing, and manipulating fractionally charged excitations of fractional chern insulators in optical lattices," *Phys. Rev. A*, vol. 98, p. 063621, Dec 2018.

[42] B. d. ż. Jaworowski, N. Regnault, and Z. Liu, "Characterization of quasiholes in two-component fractional quantum hall states and fractional chern insulators in $|c|=2$ flat bands," *Phys. Rev. B*, vol. 99, p. 045136, Jan 2019.

[43] S. Manna, J. Wildeboer, and A. E. B. Nielsen, "Quasielectrons in lattice moore-read models," *Phys. Rev. B*, vol. 99, p. 045147, Jan 2019.

[44] Z. Liu, R. N. Bhatt, and N. Regnault, "Characterization of quasiholes in fractional chern insulators," *Phys. Rev. B*, vol. 91, p. 045126, Jan 2015.

[45] F. D. M. Haldane, "Exact Jastrow-Gutzwiller resonating-valence-bond ground state of the spin-1/2 antiferromagnetic Heisenberg chain with $1/r^2$ exchange," *Phys. Rev. Lett.*, vol. 60, pp. 635–638, Feb 1988.

[46] B. S. Shastry, "Exact solution of an $S = 1/2$ Heisenberg antiferromagnetic chain with long-ranged interactions," *Phys. Rev. Lett.*, vol. 60, pp. 639–642, Feb 1988.

[47] F. D. M. Haldane, Z. N. C. Ha, J. C. Talstra, D. Bernard, and V. Pasquier, "Yangian symmetry of integrable quantum chains with long-range interactions and a new description of states in conformal field theory," *Phys. Rev. Lett.*, vol. 69, pp. 2021–2025, Oct 1992.

[48] Z. N. C. Ha and F. D. M. Haldane, "Squeezed strings and yangian symmetry of the heisenberg chain with long-range interaction," *Phys. Rev. B*, vol. 47, pp. 12459–12469, May 1993.

[49] J. C. Talstra, "Integrability and applications of the exactly-solvable haldane-shastry one-dimensional quantum spin chain," *arXiv preprint cond-mat/9509178*, 1995.

[50] J. C. Talstra and F. D. M. Haldane, "Integrals of motion of the haldane-shastry model," *Journal of Physics A: Mathematical and General*, vol. 28, pp. 2369–2377, apr 1995.

[51] F. D. M. Haldane, ""fractional statistics" in arbitrary dimensions: A generalization of the pauli principle," *Phys. Rev. Lett.*, vol. 67, pp. 937–940, Aug 1991.

[52] M. Greiter, "Statistical phases and momentum spacings for one-dimensional anyons," *Phys. Rev. B*, vol. 79, p. 064409, Feb 2009.

[53] M. Greiter, F. D. M. Haldane, and R. Thomale, "Non-abelian statistics in one dimension: Topological momentum spacings and su(2) level-k fusion rules," *Phys. Rev. B*, vol. 100, p. 115107, Sep 2019.

[54] I. Affleck, "Critical behavior of two-dimensional systems with continuous symmetries," *Phys. Rev. Lett.*, vol. 55, pp. 1355–1358, Sep 1985.

[55] F. D. M. Haldane and M. R. Zirnbauer, "Exact calculation of the ground-state dynamical spin correlation function of a $s = 1/2$ antiferromagnetic heisenberg chain with free spinons," *Phys. Rev. Lett.*, vol. 71, pp. 4055–4058, Dec 1993.

[56] L. D'Alessio, Y. Kafri, A. Polkovnikov, and M. Rigol, "From quantum chaos and eigenstate thermalization to statistical mechanics and thermodynamics," *Advances in Physics*, vol. 65, no. 3, pp. 239–362, 2016.

[57] D. N. Page, "Average entropy of a subsystem," *Phys. Rev. Lett.*, vol. 71, pp. 1291–1294, Aug 1993.

[58] P. W. Anderson, "Absence of diffusion in certain random lattices," *Phys. Rev.*, vol. 109, pp. 1492–1505, Mar 1958.

[59] D. Basko, I. Aleiner, and B. Altshuler, "Metal-insulator transition in a weakly interacting many-electron system with localized single-particle states," *Annals of Physics*, vol. 321, no. 5, pp. 1126–1205, 2006.

[60] D. J. Luitz, N. Laflorencie, and F. Alet, "Many-body localization edge in the random-field heisenberg chain," *Phys. Rev. B*, vol. 91, p. 081103, Feb 2015.

[61] M. Serbyn, Z. Papić, and D. A. Abanin, "Local conservation laws and the structure of the many-body localized states," *Phys. Rev. Lett.*, vol. 111, p. 127201, Sep 2013.

[62] A. L. Burin, "Localization in a random xy model with long-range interactions: Intermediate case between single-particle and many-body problems," *Phys. Rev. B*, vol. 92, p. 104428, Sep 2015.

[63] D. B. Gutman, I. V. Protopopov, A. L. Burin, I. V. Gornyi, R. A. Santos, and A. D. Mirlin, "Energy transport in the anderson insulator," *Phys. Rev. B*, vol. 93, p. 245427, Jun 2016.

[64] N. Y. Yao, C. R. Laumann, S. Gopalakrishnan, M. Knap, M. Müller, E. A. Demler, and M. D. Lukin, "Many-body localization in dipolar systems," *Phys. Rev. Lett.*, vol. 113, p. 243002, Dec 2014.

[65] R. M. Nandkishore and S. L. Sondhi, "Many-body localization with long-range interactions," *Phys. Rev. X*, vol. 7, p. 041021, Oct 2017.

[66] N. D. Mermin and H. Wagner, "Absence of ferromagnetism or antiferromagnetism in one- or two-dimensional isotropic heisenberg models," *Phys. Rev. Lett.*, vol. 17, pp. 1133–1136, Nov 1966.

[67] D. A. Huse, R. Nandkishore, V. Oganesyan, A. Pal, and S. L. Sondhi, "Localization-protected quantum order," *Phys. Rev. B*, vol. 88, p. 014206, Jul 2013.

[68] J. A. Kjäll, J. H. Bardarson, and F. Pollmann, "Many-body localization in a disordered quantum ising chain," *Phys. Rev. Lett.*, vol. 113, p. 107204, Sep 2014.

[69] S. A. Parameswaran and R. Vasseur, "Many-body localization, symmetry and topology," *Reports on Progress in Physics*, vol. 81, p. 082501, jul 2018.

[70] R. Vasseur, A. C. Potter, and S. A. Parameswaran, "Quantum criticality of hot random spin chains," *Phys. Rev. Lett.*, vol. 114, p. 217201, May 2015.

[71] I. V. Protopopov, W. W. Ho, and D. A. Abanin, "Effect of su(2) symmetry on many-body localization and thermalization," *Phys. Rev. B*, vol. 96, p. 041122(R), Jul 2017.

[72] I. V. Protopopov, R. K. Panda, T. Parolini, A. Scardicchio, E. Demler, and D. A. Abanin, "Non-abelian symmetries and disorder: A broad nonergodic regime and anomalous thermalization," *Phys. Rev. X*, vol. 10, p. 011025, Feb 2020.

[73] H. Bernien, S. Schwartz, A. Keesling, H. Levine, A. Omran, H. Pichler, S. Choi, A. S. Zibrov, M. Endres, M. Greiner, V. Vuletić, and M. D. Lukin, "Probing many-body dynamics on a 51-atom quantum simulator," *Nature*, vol. 551, pp. 579–584, Nov 2017.

[74] C. J. Turner, A. A. Michailidis, D. A. Abanin, M. Serbyn, and Z. Papić, "Weak ergodicity breaking from quantum many-body scars," *Nature Physics*, vol. 14, pp. 745–749, Jul 2018.

[75] N. Shiraishi and T. Mori, "Systematic construction of counterexamples to the eigenstate thermalization hypothesis," *Phys. Rev. Lett.*, vol. 119, p. 030601, Jul 2017.

[76] S. Moudgalya, N. Regnault, and B. A. Bernevig, "Entanglement of exact excited states of affleck-kennedy-lieb-tasaki models: Exact results, many-body scars, and violation of the strong eigenstate thermalization hypothesis," *Phys. Rev. B*, vol. 98, p. 235156, Dec 2018.

[77] M. Schecter and T. Iadecola, "Weak ergodicity breaking and quantum many-body scars in spin-1 xy magnets," *Phys. Rev. Lett.*, vol. 123, p. 147201, Oct 2019.

[78] T. Iadecola and M. Schecter, "Quantum many-body scar states with emergent kinetic constraints and finite-entanglement revivals," *Phys. Rev. B*, vol. 101, p. 024306, Jan 2020.

[79] S. Chattopadhyay, H. Pichler, M. D. Lukin, and W. W. Ho, "Quantum many-body scars from virtual entangled pairs," *Phys. Rev. B*, vol. 101, p. 174308, May 2020.

[80] N. Shibata, N. Yoshioka, and H. Katsura, "Onsager's scars in disordered spin chains," *Phys. Rev. Lett.*, vol. 124, p. 180604, May 2020.

[81] S. Moudgalya, B. A. Bernevig, and N. Regnault, "Quantum many-body scars in a landau level on a thin torus," *Phys. Rev. B*, vol. 102, p. 195150, Nov 2020.

[82] A. Hudomal, I. Vasić, N. Regnault, and Z. Papić, "Quantum scars of bosons with correlated hopping," *Communications Physics*, vol. 3, p. 99, Jun 2020.

[83] H. Zhao, J. Vovrosh, F. Mintert, and J. Knolle, "Quantum many-body scars in optical lattices," *Phys. Rev. Lett.*, vol. 124, p. 160604, Apr 2020.

[84] S. Ok, K. Choo, C. Mudry, C. Castelnovo, C. Chamon, and T. Neupert, "Topological many-body scar states in dimensions one, two, and three," *Phys. Rev. Research*, vol. 1, p. 033144, Dec 2019.

[85] N. S. Srivatsa, R. Moessner, and A. E. B. Nielsen, "Many-body delocalization via emergent symmetry," *Phys. Rev. Lett.*, vol. 125, p. 240401, Dec 2020.

[86] J. I. Cirac and G. Sierra, "Infinite matrix product states, conformal field theory, and the haldane-shastry model," *Phys. Rev. B*, vol. 81, p. 104431, Mar 2010.

[87] A. E. B. Nielsen, J. I. Cirac, and G. Sierra, "Quantum spin Hamiltonians for the $SU(2)_k$ WZW model," *Journal of Statistical Mechanics: Theory and Experiment*, vol. 2011, no. 11, p. P11014, 2011.

[88] M. V. Berry and M. Robnik, "Statistics of energy levels without time-reversal symmetry: Aharonov-bohm chaotic billiards," *Journal of Physics A: Mathematical and General*, vol. 19, pp. 649–668, apr 1986.

[89] V. Oganesyan and D. A. Huse, "Localization of interacting fermions at high temperature," *Phys. Rev. B*, vol. 75, p. 155111, Apr 2007.

[90] S. Pai, N. S. Srivatsa, and A. E. B. Nielsen, "Disordered haldane-shastry model," *Phys. Rev. B*, vol. 102, p. 035117, Jul 2020.

[91] F. Finkel and A. González-López, "Global properties of the spectrum of the haldane-shastry spin chain," *Phys. Rev. B*, vol. 72, p. 174411, Nov 2005.

[92] T. C. Hsu and J. C. Angle's d'Auriac, "Level repulsion in integrable and almost-integrable quantum spin models," *Phys. Rev. B*, vol. 47, pp. 14291–14296, Jun 1993.

[93] B. Herwerth, G. Sierra, H.-H. Tu, and A. E. B. Nielsen, "Excited states in spin chains from conformal blocks," *Phys. Rev. B*, vol. 91, p. 235121, Jun 2015.

[94] R. Singh, R. Moessner, and D. Roy, "Effect of long-range hopping and interactions on entanglement dynamics and many-body localization," *Phys. Rev. B*, vol. 95, p. 094205, Mar 2017.

[95] S. Nag and A. Garg, "Many-body localization in the presence of long-range interactions and long-range hopping," *Phys. Rev. B*, vol. 99, p. 224203, Jun 2019.

[96] N. S. Srivatsa, J. Wildeboer, A. Seidel, and A. E. B. Nielsen, "Quantum many-body scars with chiral topological order in two dimensions and critical properties in one dimension," *Phys. Rev. B*, vol. 102, p. 235106, Dec 2020.

[97] J. Wildeboer, A. Seidel, N. S., Srivatsa, A. E. B. Nielsen, and O. Erten, "Topological quantum many-body scars in quantum dimer models on the kagome lattice," 2020.

[98] E. P. Wigner, "Characteristic vectors of bordered matrices with infinite dimensions," *Annals of Mathematics*, vol. 62, no. 3, pp. 548–564, 1955.

[99] L. D'Alessio, Y. Kafri, A. Polkovnikov, and M. Rigol, "From quantum chaos and eigenstate thermalization to statistical mechanics and thermodynamics," *Advances in Physics*, vol. 65, no. 3, pp. 239–362, 2016.

[100] T. Takaishi, K. Sakakibara, I. Ichinose, and T. Matsui, "Localization and delocalization of fermions in a background of correlated spins," *Phys. Rev. B*, vol. 98, p. 184204, Nov 2018.

[101] A. E. B. Nielsen, "Anyon braiding in semianalytical fractional quantum hall lattice models," *Phys. Rev. B*, vol. 91, p. 041106(R), Jan 2015.

[102] A. E. B. Nielsen, I. Glasser, and I. D. Rodríguez, "Quasielectrons as inverse quasiholes in lattice fractional quantum hall models," *New Journal of Physics*, vol. 20, p. 033029, mar 2018.

[103] D. S. Rokhsar and S. A. Kivelson, "Superconductivity and the quantum hard-core dimer gas," *Phys. Rev. Lett.*, vol. 61, pp. 2376–2379, Nov 1988.

[104] G. Misguich, D. Serban, and V. Pasquier, "Quantum dimer model on the kagome lattice: Solvable dimer-liquid and ising gauge theory," *Phys. Rev. Lett.*, vol. 89, p. 137202, Sep 2002.

[105] A. Seidel, "Linear independence of nearest-neighbor valence-bond states on the kagome lattice and construction of $su(2)$-invariant spin-$\frac{1}{2}$ hamiltonian with a sutherland-rokhsar-kivelson quantum liquid ground state," *Phys. Rev. B*, vol. 80, p. 165131, Oct 2009.

[106] J. Wildeboer and A. Seidel, "Correlation functions in su(2)-invariant resonating-valence-bond spin liquids on nonbipartite lattices," *Phys. Rev. Lett.*, vol. 109, p. 147208, Oct 2012.

[107] J. Wildeboer, A. Seidel, and R. G. Melko, "Entanglement entropy and topological order in resonating valence-bond quantum spin liquids," *Phys. Rev. B*, vol. 95, p. 100402, Mar 2017.

[108] J.-Y. Chen, L. Vanderstraeten, S. Capponi, and D. Poilblanc, "Non-abelian chiral spin liquid in a quantum antiferromagnet revealed by an ipeps study," *Phys. Rev. B*, vol. 98, p. 184409, Nov 2018.

[109] D. K. Nandy, N. S. Srivatsa, and A. E. B. Nielsen, "Truncation of lattice fractional quantum hall hamiltonians derived from conformal field theory," *Phys. Rev. B*, vol. 100, p. 035123, Jul 2019.

[110] D. K. Nandy, N. S, Srivatsa, and A. E. B. Nielsen, "Local hamiltonians for one-dimensional critical models," *Journal of Statistical Mechanics: Theory and Experiment*, vol. 2018, p. 063107, jun 2018.

[111] I. Glasser, J. I. Cirac, G. Sierra, and A. E. B. Nielsen, "Exact parent hamiltonians of bosonic and fermionic moore–read states on lattices and local models," *New Journal of Physics*, vol. 17, p. 082001, aug 2015.

[112] M. Levin and X.-G. Wen, "Detecting topological order in a ground state wave function," *Phys. Rev. Lett.*, vol. 96, p. 110405, Mar 2006.

[113] F. Calogero, "Ground state of a one-dimensional n-body system," *Journal of Mathematical Physics*, vol. 10, no. 12, pp. 2197–2200, 1969.

[114] B. Sutherland, "Exact results for a quantum many-body problem in one dimension," *Phys. Rev. A*, vol. 4, pp. 2019–2021, Nov 1971.

[115] D. F. Wang, J. T. Liu, and P. Coleman, "Spectrum and thermodynamics of the one-dimensional supersymmetric t-j model with $1/r^2$ exchange and hopping," *Phys. Rev. B*, vol. 46, pp. 6639–6642, Sep 1992.

[116] R. Thomale, S. Rachel, P. Schmitteckert, and M. Greiter, "Family of spin-s chain representations of su(2)$_k$ wess-zumino-witten models," *Phys. Rev. B*, vol. 85, p. 195149, May 2012.

[117] S. M. Pittman, M. Beau, M. Olshanii, and A. del Campo, "Truncated calogero-sutherland models," *Phys. Rev. B*, vol. 95, p. 205135, May 2017.

[118] C. Gross and I. Bloch, "Quantum simulations with ultracold atoms in optical lattices," *Science*, vol. 357, no. 6355, pp. 995–1001, 2017.

[119] C. Trefzger, C. Menotti, B. Capogrosso-Sansone, and M. Lewenstein, "Ultracold dipolar gases in optical lattices," *Journal of Physics B: Atomic, Molecular and Optical Physics*, vol. 44, p. 193001, sep 2011.

[120] T. Sowiński, O. Dutta, P. Hauke, L. Tagliacozzo, and M. Lewenstein, "Dipolar molecules in optical lattices," *Phys. Rev. Lett.*, vol. 108, p. 115301, Mar 2012.

[121] S. Baier, M. J. Mark, D. Petter, K. Aikawa, L. Chomaz, Z. Cai, M. Baranov, P. Zoller, and F. Ferlaino, "Extended bose-hubbard models with ultracold magnetic atoms," *Science*, vol. 352, no. 6282, pp. 201–205, 2016.

[122] S. Greschner, L. Santos, and T. Vekua, "Ultracold bosons in zig-zag optical lattices," *Phys. Rev. A*, vol. 87, p. 033609, Mar 2013.

[123] A. Dhar, T. Mishra, R. V. Pai, S. Mukerjee, and B. P. Das, "Hard-core bosons in a zig-zag optical superlattice," *Phys. Rev. A*, vol. 88, p. 053625, Nov 2013.

[124] T. Zhang and G.-B. Jo, "One-dimensional sawtooth and zigzag lattices for ultracold atoms," *Scientific Reports*, vol. 5, p. 16044, Nov 2015.

[125] E. Anisimovas, M. Račiūnas, C. Sträter, A. Eckardt, I. B. Spielman, and G. Juzeliūnas, "Semisynthetic zigzag optical lattice for ultracold bosons," *Phys. Rev. A*, vol. 94, p. 063632, Dec 2016.

[126] P. Calabrese and J. Cardy, "Entanglement entropy and quantum field theory," *Journal of Statistical Mechanics: Theory and Experiment*, vol. 2004, p. P06002, jun 2004.

[127] N. S. Srivatsa, X. Li, and A. E. B. Nielsen, "Squeezing anyons for braiding on small lattices," *Phys. Rev. Research*, vol. 3, p. 033044, Jul 2021.

[128] A. E. B. Nielsen, I. Glasser, and I. D. Rodríguez, "Quasielectrons as inverse quasiholes in lattice fractional quantum Hall models," *New Journal of Physics*, vol. 20, p. 033029, mar 2018.

[129] E. Kapit and E. Mueller, "Exact parent hamiltonian for the quantum Hall states in a lattice," *Phys. Rev. Lett.*, vol. 105, p. 215303, Nov 2010.

[130] N. Hansen, *The CMA Evolution Strategy: A Comparing Review*, pp. 75–102. Springer-Verlag Berlin Heidelberg, 2006.

[131] X. Chen, Z.-C. Gu, and X.-G. Wen, "Local unitary transformation, long-range quantum entanglement, wave function renormalization, and topological order," *Phys. Rev. B*, vol. 82, p. 155138, Oct 2010.

[132] X. G. Wen and Q. Niu, "Ground-state degeneracy of the fractional quantum Hall states in the presence of a random potential and on high-genus Riemann surfaces," *Phys. Rev. B*, vol. 41, pp. 9377–9396, May 1990.

[133] R. Tao and F. D. M. Haldane, "Impurity effect, degeneracy, and topological invariant in the quantum Hall effect," *Phys. Rev. B*, vol. 33, pp. 3844–3850, Mar 1986.

[134] Q. Niu, D. J. Thouless, and Y.-S. Wu, "Quantized Hall conductance as a topological invariant," *Phys. Rev. B*, vol. 31, pp. 3372–3377, Mar 1985.

[135] K. Kudo, H. Watanabe, T. Kariyado, and Y. Hatsugai, "Many-body Chern number without integration," *Phys. Rev. Lett.*, vol. 122, p. 146601, Apr 2019.

[136] T. Neupert, L. Santos, C. Chamon, and C. Mudry, "Fractional quantum Hall states at zero magnetic field," *Phys. Rev. Lett.*, vol. 106, p. 236804, Jun 2011.

[137] W.-J. Hu, S.-S. Gong, W. Zhu, and D. N. Sheng, "Competing spin-liquid states in the spin-$\frac{1}{2}$ Heisenberg model on the triangular lattice," *Phys. Rev. B*, vol. 92, p. 140403(R), Oct 2015.

[138] N. Regnault and B. A. Bernevig, "Fractional Chern insulator," *Phys. Rev. X*, vol. 1, p. 021014, Dec 2011.

[139] R. Thomale, A. Sterdyniak, N. Regnault, and B. A. Bernevig, "Entanglement gap and a new principle of adiabatic continuity," *Phys. Rev. Lett.*, vol. 104, p. 180502, May 2010.

[140] M. Hermanns, A. Chandran, N. Regnault, and B. A. Bernevig, "Haldane statistics in the finite-size entanglement spectra of $1/m$ fractional quantum Hall states," *Phys. Rev. B*, vol. 84, p. 121309(R), Sep 2011.

[141] A. Sterdyniak, N. Regnault, and B. A. Bernevig, "Extracting excitations from model state entanglement," *Phys. Rev. Lett.*, vol. 106, p. 100405, Mar 2011.

[142] M. Levin and X.-G. Wen, "Detecting topological order in a ground state wave function," *Phys. Rev. Lett.*, vol. 96, p. 110405, Mar 2006.

[143] A. Kitaev and J. Preskill, "Topological entanglement entropy," *Phys. Rev. Lett.*, vol. 96, p. 110404, Mar 2006.

[144] H.-C. Jiang, Z. Wang, and L. Balents, "Identifying topological order by entanglement entropy," *Nature Physics*, vol. 8, pp. 902–905, 2012.

[145] P. Zanardi and N. Paunković, "Ground state overlap and quantum phase transitions," *Phys. Rev. E*, vol. 74, p. 031123, Sep 2006.

[146] S. Manna, N. S. Srivatsa, J. Wildeboer, and A. E. B. Nielsen, "Quasiparticles as detector of topological quantum phase transitions," *Phys. Rev. Research*, vol. 2, p. 043443, Dec 2020.

[147] G. Moore and N. Read, "Nonabelions in the fractional quantum Hall effect," *Nuclear Physics B*, vol. 360, no. 2, pp. 362–396, 1991.

[148] P. Bonderson, V. Gurarie, and C. Nayak, "Plasma analogy and non-Abelian statistics for Ising-type quantum Hall states," *Phys. Rev. B*, vol. 83, p. 075303, Feb 2011.

[149] I. Glasser, J. I. Cirac, G. Sierra, and A. E. B. Nielsen, "Exact parent Hamiltonians of bosonic and fermionic Moore-Read states on lattices and local models," *New Journal of Physics*, vol. 17, no. 8, p. 082001, 2015.

[150] S. Manna, J. Wildeboer, and A. E. B. Nielsen, "Quasielectrons in lattice Moore-Read models," *Phys. Rev. B*, vol. 99, p. 045147, Jan 2019.

[151] S. Manna, B. Pal, W. Wang, and A. E. B. Nielsen, "Anyons and fractional quantum hall effect in fractal dimensions," *Phys. Rev. Research*, vol. 2, p. 023401, Jun 2020.

[152] M. Hafezi, A. S. Sørensen, E. Demler, and M. D. Lukin, "Fractional quantum Hall effect in optical lattices," *Phys. Rev. A*, vol. 76, p. 023613, Aug 2007.

[153] A. S. Sørensen, E. Demler, and M. D. Lukin, "Fractional quantum Hall states of atoms in optical lattices," *Phys. Rev. Lett.*, vol. 94, p. 086803, Mar 2005.

[154] E. Kapit, P. Ginsparg, and E. Mueller, "Non-Abelian braiding of lattice bosons," *Phys. Rev. Lett.*, vol. 108, p. 066802, Feb 2012.

[155] A. E. B. Nielsen, I. Glasser, and I. D. Rodríguez, "Quasielectrons as inverse quasiholes in lattice fractional quantum Hall models," *New Journal of Physics*, vol. 20, no. 3, p. 033029, 2018.

[156] A. Kitaev, "Anyons in an exactly solved model and beyond," *Annals of Physics*, vol. 321, pp. 2–111, 2006.

[157] A. Kitaev, "Fault-tolerant quantum computation by anyons," *Annals of Physics*, vol. 303, pp. 2–30, 2003.

[158] J. Vidal, S. Dusuel, and K. P. Schmidt, "Low-energy effective theory of the toric code model in a parallel magnetic field," *Phys. Rev. B*, vol. 79, p. 033109, Jan 2009.

[159] J. Vidal, R. Thomale, K. P. Schmidt, and S. Dusuel, "Self-duality and bound states of the toric code model in a transverse field," *Phys. Rev. B*, vol. 80, p. 081104(R), Aug 2009.

[160] S. Dusuel, M. Kamfor, R. Orús, K. P. Schmidt, and J. Vidal, "Robustness of a perturbed topological phase," *Phys. Rev. Lett.*, vol. 106, p. 107203, Mar 2011.

[161] M. H. Zarei, "Ising order parameter and topological phase transitions: Toric code in a uniform magnetic field," *Phys. Rev. B*, vol. 100, p. 125159, Sep 2019.

[162] E. Greplova, A. Valenti, G. Boschung, F. Schäfer, N. Lörch, and S. D. Huber, "Unsupervised identification of topological phase transitions using predictive models," *New Journal of Physics*, vol. 22, p. 045003, apr 2020.

[163] C. Weitenberg, M. Endres, J. F. Sherson, M. Cheneau, P. Schauß, T. Fukuhara, I. Bloch, and S. Kuhr, "Single-spin addressing in an atomic Mott insulator," *Nature*, vol. 471, no. 7338, p. 319, 2011.

[164] T. Chen, S. Zhang, Y. Zhang, Y. Liu, S.-P. Kou, H. Sun, and X. Zhang, "Experimental observation of classical analogy of topological entanglement entropy," *Nature communications*, vol. 10, no. 1, p. 1557, 2019.

[165] Z. Luo, J. Li, Z. Li, L.-Y. Hung, Y. Wan, X. Peng, and J. Du, "Experimentally probing topological order and its breakdown through modular matrices," *Nature physics*, vol. 14, pp. 160–165, 2018.

[166] J. K. Pachos, W. Wieczorek, C. Schmid, N. Kiesel, R. Pohlner, and H. Weinfurter, "Revealing anyonic features in a toric code quantum simulation," *New Journal of Physics*, vol. 11, no. 8, p. 083010, 2009.

[167] P. Francesco, P. Mathieu, and D. Sénéchal. Graduate Texts in Contemporary Physics, Springer-Verlag New York.

[168] A. Deshpande and A. E. B. Nielsen, "Lattice laughlin states on the torus from conformal field theory," *Journal of Statistical Mechanics: Theory and Experiment*, vol. 2016, p. 013102, jan 2016.

www.ingramcontent.com/pod-product-compliance
Lightning Source LLC
LaVergne TN
LVHW042202190726
843493LV00006B/1780